Engineering Economic Analysis Exam File

Engineering Economic Analysis Exam File

Donald G. Newnan, Editor
Hal Ball, Umpqua College
B. G. Barr, The University of Kansas
Charles Burford, Texas Tech University
Gary Crossman, Old Dominion University
Anthony DeFruscio, University of Lowell
R. James Diegel, Lawrence Institute of Technology
John Eastman, Michigan State University
Bryan Jenkins, University of California, Davis
Jack Lohmann, The University of Michigan
Hugo Patino, University of California, Davis
Joseph Pignatiello, Jr., University of Arizona
Paul Schonfeld, University of Maryland
Barry Shiller, San Francisco State University
Ralph Smith, University of Georgia
Thomas Ward, University of Louisville

Engineering Press, Inc. San Jose, California

Library of Congress Cataloging-in-Publication Data

Engineering economic analysis exam file / Donald G. Newnan, editor; Hal Ball . . . [et al.].
p. cm. – (Exam File series)
ISBN 0-910554-81-1
1. Engineering economy—Problems, exercises, etc. I. Newnan, Donald G.
II. Ball, Hal. Series.
TA177.4.E535 1991
658.1'5—dc20 90-3901
CIP

Engineering economic analysis.
Fourth Edition: ISBN 0-910554-82-X
Exam File: ISBN 0-910554-81-1
The two volumes are not sold separately.
They are shrinkwrapped and sold only as a set.
Order ISBN 0-910554-83-8.

Cover illustration courtesy of Monex International Ltd.
Printed in the United States of America 4 3 2 1

Engineering Press, Inc. **P.O. Box 1** **San Jose, California 95103-0001**

Contents

Foreword vii

Course Summary 1

Professors' Examination Problems and Solutions 33

1 Introductory Problems—Basic Concepts 35

Decision-making without compound interest tables

2 Interest Problems and Equivalence 43

Using compound interest tables
Periodic compounding and continuous compounding
Nominal and effective interest

3 Present Worth 65

Present Worth of future sums of money
Choosing between alternatives
Capitalized Cost (PW of infinite series)

4 Annual Cash Flow 89

Annual Equivalent of an irregular cash flow
Equivalent Uniform Annual Cost (EUAC)
Choosing between alternatives

5 Computing an Unknown Interest Rate 119

6 Rate of Return Analysis 127

Solving two alternative problems
Incremental analysis
Solving multiple alternative problems

7 Other Analysis Techniques 151

Future Worth
Payback (payoff) Period
Sensitivity and breakeven

8 Depreciation Computations 171

Straight Line and Sum-Of-Years-Digits
Declining Balance and Double Declining Balance
Accelerated Cost Recovery System (ACRS)

9 Income Taxes 189

Personal and corporate income taxes
After-tax economic analysis

10 Equipment Retirement and Replacement 219

11 Inflation 231

12 Benefit-Cost Ratio 249

Economic analysis in government

13 Other Problems 257

Compound Interest Tables 265

Foreword

As an engineering professor, I have observed that students often do not have useful study materials. They usually have only the course textbook and their own lecture notes and, having seen many students' lecture notes, these can leave much to be desired in terms of clear, organized information.

On the other hand, professors usually save their written examinations and solutions to problems posed in tests and homework. Sometimes these are used in subsequent classes, but mostly they are filed away. Many instructors reason that the old exams can be employed when a new test must be prepared quickly. Unfortunately, however, much of this excellent study material is lost to the very people we are endeavoring to teach.

This volume seeks to remedy this problem. Professors from around the country have opened their exam files and allowed their problems and solutions to be published. These professors all teach an introductory course in engineering economic analysis/engineering economy and use one of the half-dozen popular textbooks. The problems were carefully selected for publication so that the fundamentals of engineering economic analysis/engineering economy found in these textbooks would be covered.

In general, the problems are just as they appeared in an actual course examination. Similarly, the solutions were prepared by the professor who wrote the exam problem.

This book begins with a summary of a typical engineering economic analysis/engineering economy course. The main section of the book consists of 384 authentic examination problems with the professors' own solutions. The last section has a set of compound interest tables.

The large task of transcribing the professors' solutions into readable presentations was ably performed by Bonnie Leong. If your working of any problem uncovers an error in the statement or solution, kindly inform the editor at the Engineering Press address. I hope you find this material facilitates a better understanding of engineering economic analysis/engineering economy and that it helps you to succeed on your course examinations!

Donald G. Newnan
Editor

Course Summary

This chapter is a brief review of engineering economic analysis/engineering economy. The goal is to give you a better grasp of the major topics in a typical first course. Hopefully this overview will help you put the course lectures and your reading of the textbook in better perspective. There are 26 example problems scattered throughout the engineering economics review. These examples are an integral part of the review and should be worked to completion as you come to them.

CASH FLOW

The field of engineering economics uses mathematical and economic techniques to systematically analyze situations which pose alternative courses of action.

The initial step in engineering economics problems is to resolve a situation, or each alternative course in a given situation, into its favorable and unfavorable consequences or factors. These are then measured in some common unit—usually money. Those factors which cannot readily be reduced to money are called intangible, or irreducible, factors. Intangible or irreducible factors are not included in any monetary analysis but are considered in conjunction with such an analysis when making the final decision on proposed courses of action.

A cash flow table shows the "money consequences" of a situation and its timing. For example, a simple problem might be to list the year-by-year consequences of purchasing and owning a used car:

Year	*Cash Flow*	
Beginning of first Year 0	−$4500	Car purchased "now" for $4500 cash. The minus sign indicates a disbursement.
End of Year 1	−350	Maintenance costs are $350 per year.
End of Year 2	−350	
End of Year 3	−350	
End of Year 4	−350	
	+2000	The car is sold at the end of the 4th year for $2000. The plus sign represents a receipt of money.

This same cash flow may be represented graphically:

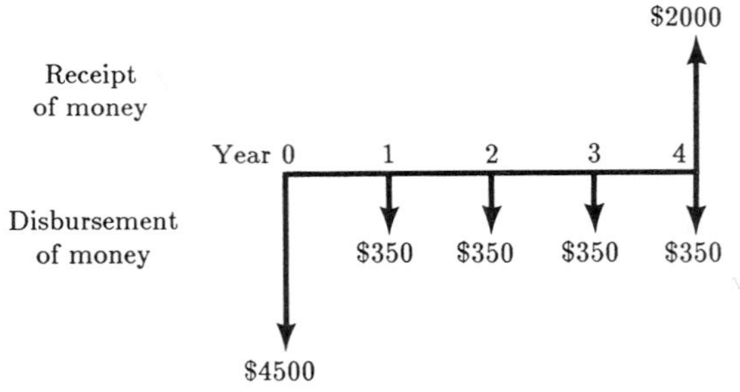

The upward arrow represents a receipt of money, and the downward arrows represent disbursements. The x-axis represents the passage of time.

EXAMPLE 1

In January 1990, a firm purchases a used typewriter for $500. Repairs cost nothing in 1990 or 1991. Repairs are $85 in 1992, $130 in 1993, and $140 in 1994. The machine is sold in 1994 for $300. Compute the cash flow table.

Solution

Unless otherwise stated in problems, the customary assumption is a beginning-of-year purchase, followed by end-of-year receipts or disbursements, and an end-of-year resale or salvage value. Thus the typewriter repairs and the typewriter sale are assumed to occur at the end of the year. Letting a minus sign represent a disbursement of money, and a plus sign a receipt of money, we are able to set up this cash flow table:

Year	*Cash Flow*
Beginning of 1990	−$500
End of 1990	0
End of 1991	0
End of 1992	−85
End of 1993	−130
End of 1994	+160

Notice that at the end of 1994, the cash flow table shows +160 which is the net sum of −140 and +300. If we define Year 0 as the beginning of 1990, the cash flow table becomes:

Year	*Cash Flow*
0	−$500
1	0
2	0
3	−85
4	−130
5	+160

From this cash flow table, the definitions of Year 0 and Year 1 become clear. Year 0 is defined as the *beginning* of Year 1. Year 1 is the *end* of Year 1. Year 2 is the *end* of Year 2, and so forth.

TIME VALUE OF MONEY

When the money consequences of an alternative occur in a short period of time—say, less than one year—we might simply add up the various sums of money and obtain the net result. But we cannot treat money this same way over longer periods of time. This is because money today is not the same as money at some future time.

Consider this question: Which would you prefer, $100 today or the assurance of receiving $100 a year from now? Clearly, you would prefer the $100 today. If you had the money today, rather than a year from now, you could use it for the year. And if you had no use for it, you could lend it to someone who would pay interest for the privilege of using your money for the year.

EQUIVALENCE

In the preceding section we saw that money at different points in time (for example, $100 today or $100 one year hence) may be equal in the sense that they both are $100, but $100 a year hence is *not* an acceptable substitute for $100 today. When we have acceptable substitutes, we say they are *equivalent* to each other. Thus at 8% interest, $108 a year hence is equivalent to $100 today.

EXAMPLE 2

At a 10% per year interest rate, $500 now is *equivalent* to how much three years hence?

Solution

$500 now will increase by 10% in each of the three years.

Now =			$500
End of 1st year =	500 + 10%(500)	=	550
End of 2nd year =	550 + 10%(550)	=	605
End of 3rd year =	605 + 10%(605)	=	665.50

Thus $500 now is *equivalent* to $665.50 at the end of three years.

Equivalence is an essential factor in engineering economic analysis. Suppose we wish to select the better of two alternatives. First, we must compute their cash flows. An example would be:

	Alternative	
Year	*A*	*B*
0	−\$2000	−\$2800
1	+800	+1100
2	+800	+1100
3	+800	+1100

The larger investment in Alternative *B* results in larger subsequent benefits, but we have no direct way of knowing if Alternative *B* is better than Alternative *A*. Therefore we do not know which alternative should be selected. To make a decision we must resolve the alternatives into *equivalent* sums so they may be compared accurately and a decision made.

COMPOUND INTEREST FACTORS

To facilitate equivalence computations a series of compound interest factors will be derived and their use illustrated.

Symbols

i = Interest rate per interest period. In equations the interest rate is stated as a decimal (that is, 8% interest is 0.08).

n = Number of interest periods.

P = A present sum of money.

F = A future sum of money. The future sum F is an amount, n interest periods from the present, that is equivalent to P with interest rate i.

A = An end-of-period cash receipt or disbursement in a uniform series continuing for n periods, the entire series equivalent to P or F at interest rate i.

G = Uniform period-by-period increase in cash flows; the arithmetic gradient.

g = Uniform *rate* of period-by-period increase in cash flows; the geometric gradient.

Functional Notation

	To Find	*Given*	*Functional Notation*
Single Payment			
Compound Amount Factor	F	P	$(F/P,i,n)$
Present Worth Factor	P	F	$(P/F,i,n)$

FUNCTIONAL NOTATION, continued

	To Find	*Given*	*Functional Notation*
Uniform Payment Series			
Sinking Fund Factor	A	F	$(A/F,i,n)$
Capital Recovery Factor	A	P	$(A/P,i,n)$
Compound Amount Factor	F	A	$(F/A,i,n)$
Present Worth Factor	P	A	$(P/A,i,n)$
Arithmetic Gradient			
Gradient Uniform Series	A	G	$(A/G,i,n)$
Gradient Present Worth	P	G	$(P/G,i,n)$

From the table above we can see that the functional notation scheme is based on writing (To Find / Given,i,n). Thus, if we wished to find the future sum F, given a uniform series of receipts A, the proper compound interest factor to use would be $(F/A,i,n)$.

Single Payment Formulas

Suppose a present sum of money P is invested for one year at interest rate i. At the end of the year, we receive back our initial investment P together with interest equal to Pi or a total amount $P + Pi$. Factoring P, the sum at the end of one year is $P(1 + i)$. If we agree to let our investment remain for subsequent years, the progression is as follows:

	Amount at Beginning of Period	+	Interest for the Period	=	Amount at End of the Period
1st year	P	+	Pi	=	$P(1 + i)$
2nd year	$P(1 + i)$	+	$Pi(1 + i)$	=	$P(1 + i)^2$
3rd year	$P(1 + i)^2$	+	$Pi(1 + i)^2$	=	$P(1 + i)^3$
nth year	$P(1 + i)^{n-1}$	+	$Pi(1 + i)^{n-1}$	=	$P(1 + i)^n$

The present sum P increases in n periods to $P(1 + i)^n$. This gives us a relationship between a present sum P and its equivalent future sum F:

$$\text{Future Sum} = (\text{Present Sum})(1 + i)^n$$
$$F = P(1 + i)^n$$

This is the Single Payment Compound Amount formula. In functional notation it is written:

$$F = P(F/P,i,n)$$

The relationship may be rewritten as:

$$\text{Present Sum} = (\text{Future Sum})(1 + i)^{-n}$$
$$P = F(1 + i)^{-n}$$

This is the Single Payment Present Worth formula. It is written:

$$P = F(P/F,i,n)$$

EXAMPLE 3

At a 10% per year interest rate, \$500 now is *equivalent* to how much three years hence?

Solution

This problem was solved in Example 2. Now it can be solved using a single payment formula.

$$P = \$500$$
$$n = 3 \text{ years}$$
$$i = 10\%$$
$$F = \text{unknown}$$

$$F = P(1 + i)^n = 500(1 + 0.10)^3 = \$665.50$$

This problem may also be solved using the Compound Interest Tables.

$$F = P(F/P,i,n) = 500(F/P,10\%,3)$$

From the 10% Compound Interest Table, read $(F/P,10\%,3) = 1.331$.

$$F = 500(F/P,10\%,3) = 500(1.331) = \$665.50$$

EXAMPLE 4

To raise money for a new business, a man asks you to loan him some money. He offers to pay you \$3000 at the end of four years. How much should you give him now if you want 12% interest per year on your money?

Solution

$$P = \text{unknown}$$
$$n = 4 \text{ years}$$
$$i = 12\%$$
$$F = \$3000$$

$$P = F(1 + i)^{-n} = 3000(1 + 0.12)^{-4} = \$1906.55$$

Alternate computation using Compound Interest Tables:

$$\begin{aligned} P = F(P/F,i,n) &= 3000(P/F,12\%,4) \\ &= 3000(0.6355) = \$1906.50 \end{aligned}$$

Note that the solution based on the Compound Interest Table is slightly different from the exact solution using a hand calculator. In economic analysis, the Compound Interest Tables are always considered to be sufficiently accurate.

Uniform Payment Series Formulas

Consider the following situation:

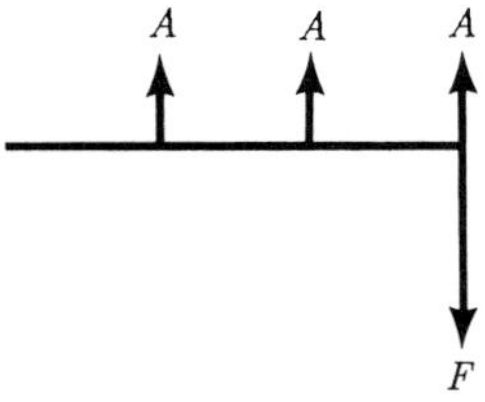

$A =$ End-of-period cash receipt or disbursement in a uniform series continuing for n periods.
$F =$ A future sum of money.

Using the single payment compound amount factor, we can write an equation for F in terms of A:

$$F = A + A(1 + i) + A(1 + i)^2 \qquad (1)$$

In our situation, with $n = 3$, Equation (1) may be written in a more general form:

$$F = A + A(1 + i) + A(1 + i)^{n-1} \qquad (2)$$

Multiply Eq. (2) by $(1 + i)$: $(1 + i)F = \cancel{A(1 + i)} + \cancel{A(1 + i)^{n-1}} + A(1 + i)^n$ (3)

Write Eq. (2): $- \quad F = A + \cancel{A(1 + i)} + \cancel{A(1 + i)^{n-1}}$ (2)

(3) − (2): $iF = -A + A(1 + i)^n$

$$F = A\left(\frac{(1 + i)^n - 1}{i}\right)$$

Uniform Series Compound Amount formula

Solving this equation for A:

$$A = F\left(\frac{i}{(1 + i)^n - 1}\right)$$ Uniform Series Sinking Fund formula

Since $F = P(1 + i)^n$, we can substitute this expression for F in the equation and obtain:

$$A = P\left(\frac{i(1 + i)^n}{(1 + i)^n - 1}\right)$$ Uniform Series Capital Recovery formula

Solving the equation for P:

$$P = A\left(\frac{(1 + i)^n - 1}{i(1 + i)^n}\right)$$ Uniform Series Present Worth formula

In functional notation, the uniform series factors are:

Compound Amount	$(F/A,i,n)$
Sinking Fund	$(A/F,i,n)$
Capital Recovery	$(A/P,i,n)$
Present Worth	$(P/A,i,n)$

EXAMPLE 5

If \$100 is deposited at the end of each year in a savings account that pays 6% interest per year, how much will be in the account at the end of five years?

Solution

A = \$100
F = unknown
n = 5 years
i = 6%

$$F = A(F/A,i,n) = 100(F/A,6\%,5)$$
$$= 100(5.637) = \$563.70$$

EXAMPLE 6

A woman wishes to make a uniform deposit every three months to her savings account so that at the end of 10 years she will have \$10,000 in the account. If the account earns 6% annual interest, compounded quarterly, how much should she deposit each three months?

Solution

F = \$10,000
A = unknown
n = 40 quarterly deposits
i = $1\frac{1}{2}\%$ per quarter year

Note that i, the interest rate per interest period, is $1\frac{1}{2}\%$, and there are 40 deposits.

$$A = F(A/F,i,n) = 10{,}000(A/F,1\tfrac{1}{2}\%,40)$$
$$= 10{,}000(0.0184) = \$184$$

EXAMPLE 7

An individual is considering the purchase of a used automobile. The total price is \$6200 with \$1240 as a downpayment and the balance paid in 48 equal monthly payments with interest at 1% per month. The payments are due at the end of each month. Compute the monthly payment.

Solution

The amount to be repaid by the 48 monthly payments is the cost of the automobile *minus* the \$1240 downpayment.

P = \$4960
A = unknown
n = 48 monthly payments
i = 1% per month

$$A = P(A/P,i,n) = 4960(A/P,1\%,48) = 4960(0.0263) = \$130.45$$

EXAMPLE 8

A couple sold their home. In addition to cash, they took a mortgage on the house. The mortgage will be paid off by monthly payments of \$232.50 for 10 years. The couple decides to sell the mortgage to a local bank. The bank will buy the mortgage, but requires a 1% per month interest rate on their investment. How much will the bank pay for the mortgage?

Solution

A = \$232.50
n = 120 months
i = 1% per month
P = unknown

$$P = A(P/A,i,n) = 232.50(P/A,1\%,120) = 232.50(69.701) = \$16,205.48$$

Arithmetic Gradient

At times one will encounter a situation where the cash flow series is not a constant amount A. Instead it is an increasing series like:

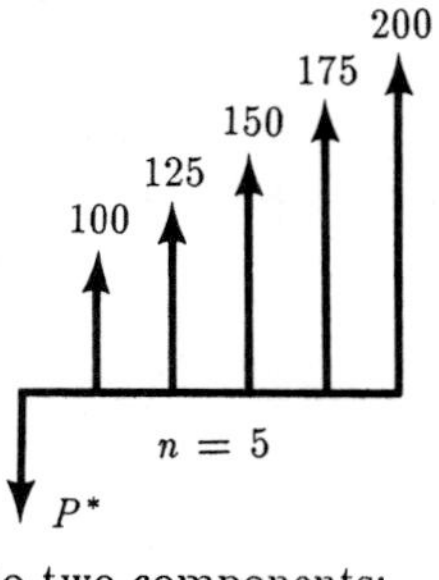

This cash flow may be resolved into two components:

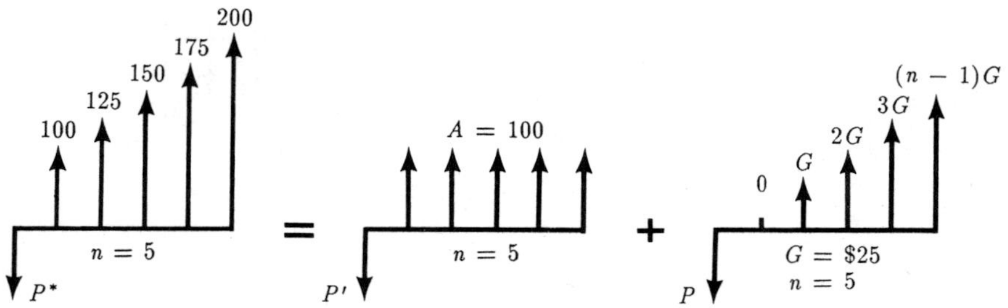

We can compute the value of P^* as equal to P' plus P. We already have an equation for P':

$$P' = A(P/A,i,n)$$

The value for P in the right-hand diagram is:

$$P = G\left(\frac{(1+i)^n - in - 1}{i^2(1+i)^n}\right)$$

This is the Arithmetic Gradient Present Worth formula. In functional notation, the relationship is $P = G(P/G,i,n)$.

EXAMPLE 9

The maintenance on a machine is expected to be \$155 at the end of the first year, and increasing \$35 each year for the following seven years. What present sum of money would need to be set aside now to pay the maintenance for the eight-year period? Assume 6% interest.

Solution

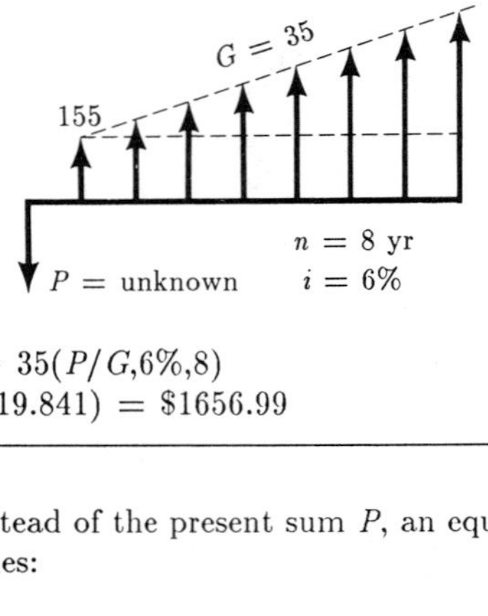

$$P = 155(P/A,6\%,8) + 35(P/G,6\%,8)$$
$$= 155(6.210) + 35(19.841) = \$1656.99$$

In the gradient series, if instead of the present sum P, an equivalent uniform series A is desired, the problem becomes:

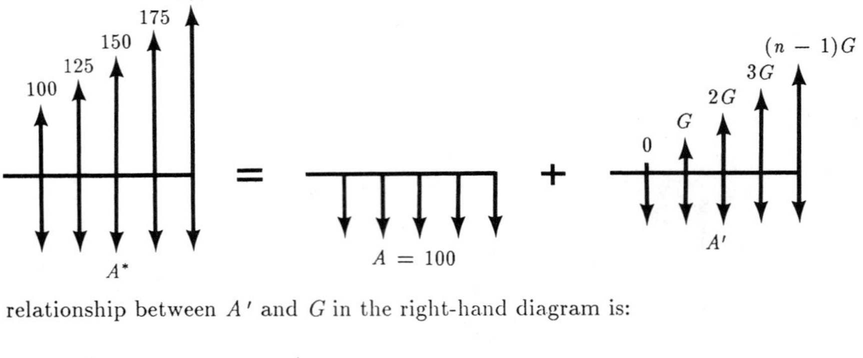

The relationship between A' and G in the right-hand diagram is:

$$A' = G\left(\frac{(1+i)^n - in - 1}{i(1+i)^n - i}\right)$$

In functional notation, the Arithmetic Gradient (to) Uniform Series factor is:

$$A = G(A/G,i,n)$$

It is important to note carefully the diagrams for the two arithmetic gradient series factors. In both cases the first term in the arithmetic gradient series is zero and the last term is $(n - 1)G$. But we use n in the equations and functional notation. The derivations (not shown here) were done on this basis and the arithmetic gradient series Compound Interest Tables are computed this way.

EXAMPLE 10

For the situation in Example 9, we wish now to know the uniform annual maintenance cost. Compute an equivalent A for the maintenance costs.

Solution

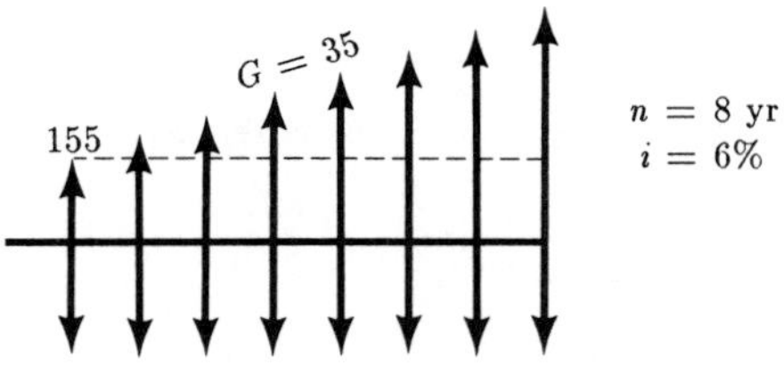

Equivalent uniform annual maintenance cost:

$$A = 155 + 35(A/G,6\%,8)$$
$$= 155 + 35(3.195) = \$266.83$$

Geometric Gradient

The arithmetic gradient is applicable where the period-by-period change in the cash flow is a uniform amount. There are other situations where the period-by-period change is a *uniform rate*, g. A diagram of this situation is:

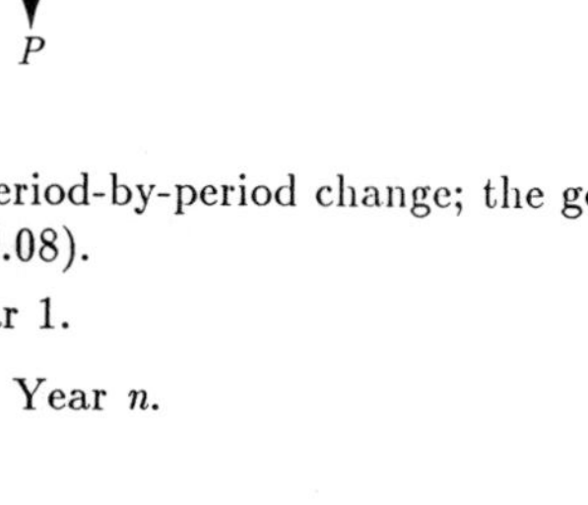

where $A_n = A_1(1 + g)^{n-1}$

g = Uniform rate of period-by-period change; the geometric gradient stated as a decimal ($8\% = 0.08$).

A_1 = Value of A at Year 1.

A_n = Value of A at any Year n.

Geometric Series Present Worth Formulas:

When $i = g$, $P = A_1\left(n(1 + i)^{-1}\right)$

When $i \neq g$, $P = A_1\left(\dfrac{1 - (1 + g)^n(1 + i)^{-n}}{i - g}\right)$

EXAMPLE 11

It is likely that airplane tickets will increase 8% in each of the next four years. The cost of a plane ticket at the end of the first year will be \$180. How much money would need to be placed in a savings account now to have money to pay a student's travel home at the end of each year for the next four years? Assume the savings account pays 5% annual interest.

Solution

The problem describes a geometric gradient where $g = 8\%$ and $i = 5\%$.

$$P = A_1\left(\frac{1 - (1 + g)^n(1 + i)^{-n}}{i - g}\right)$$

$$= 180.00\left(\frac{1 - (1.08)^4(1.05)^{-4}}{0.05 - 0.08}\right) = 180.00\left(\frac{-0.119278}{-0.03}\right) = \$715.67$$

Thus, \$715.67 would need to be deposited now.

As a check, the problem can be solved without using the geometric gradient:

Year		*Ticket*
1	$A_1 =$	\$180.00
2	$A_2 = 180.00 + 8\%(180.00) =$	194.40
3	$A_3 = 194.40 + 8\%(194.40) =$	209.95
4	$A_4 = 209.95 + 8\%(209.95) =$	226.75

$$\begin{aligned} P &= 180.00(P/F{,}5\%{,}1) + 194.40(P/F{,}5\%{,}2) + 209.95(P/F{,}5\%{,}3) + 226.75(P/F{,}5\%{,}4) \\ &= 180.00(0.9524) + 194.40(0.9070) + 209.95(0.8638) + 226.75(0.8227) \\ &= \$715.66 \end{aligned}$$

NOMINAL AND EFFECTIVE INTEREST

Nominal interest is the annual interest rate without considering the effect of any compounding.

Effective interest is the annual interest rate taking into account the effect of any compounding during the year.

Frequently an interest rate is described as an annual rate, even though the interest period may be something other than one year. A bank may pay 1–1/2% interest on the amount in a savings account every three months. The *nominal* interest rate in this situation is 6% (4 × 1–1/2% = 6%). But if you deposited $1000 in such an account, would you have 106%(1000) = $1060 in the account at the end of one year? The answer is no, you would have more. The amount in the account would increase as follows:

		Amount in Account
	At beginning of year =	$1000.00
End of 3 months:	$1000.00 + 1\frac{1}{2}\%(1000.00) =$	1015.00
End of 6 months:	$1015.00 + 1\frac{1}{2}\%(1015.00) =$	1030.23
End of 9 months:	$1030.23 + 1\frac{1}{2}\%(1030.23) =$	1045.68
End of one year:	$1045.68 + 1\frac{1}{2}\%(1045.68) =$	1061.37

The actual interest rate on the $1000 would be the interest, $61.37, divided by the original $1000, or 6.137%. We call this the *effective* interest rate.

Effective interest rate $= (1 + i)^m - 1$, where

i = Interest rate per interest period;
m = Number of compoundings per year.

EXAMPLE 12
A bank charges $1\frac{1}{2}\%$ per month on the unpaid balance for purchases made on its credit card. What nominal interest rate is it charging? What effective interest rate?

Solution
The nominal interest rate is simply the annual interest ignoring compounding, or $12(1\frac{1}{2}\%) = 18\%$.

Effective interest rate $= (1 + 0.015)^{12} - 1 = 0.1956 = 19.56\%$

SOLVING ECONOMIC ANALYSIS PROBLEMS

The techniques presented so far illustrate how to convert single amounts of money, and uniform or gradient series of money, into some equivalent sum at another point in time. These compound interest computations are an essential part of economic analysis problems.

The typical situation is that we have a number of alternatives and the question is, which alternative should be selected? The customary method of solution is to resolve each of the alternatives into some common form and then choose the best alternative (taking both the monetary and intangible factors into account).

Criteria

Economic analysis problems inevitably fall into one of three categories:

1.	Fixed Input	The amount of money or other input resources is fixed.

Example: A project engineer has a budget of $450,000 to overhaul a plant.

2.	Fixed Output	There is a fixed task, or other output to be accomplished.

Example: A mechanical contractor has been awarded a fixed price contract to air-condition a building.

3.	Neither Input nor Output Fixed	This is the general situation where neither the amount of money or other inputs, nor the amount of benefits or other outputs are fixed.

Example: A consulting engineering firm has more work available than it can handle. It is considering paying the staff for working evenings to increase the amount of design work it can perform.

There are five major methods of comparing alternatives: present worth; future worth; annual cost; rate of return; and benefit–cost ratio. These are presented in the following sections.

PRESENT WORTH

In present worth analysis, the approach is to resolve all the money consequences of an alternative into an equivalent present sum. For the three categories given above, the criteria are:

Category	*Present Worth Criterion*
Fixed Input	Maximize the Present Worth of benefits or other outputs.
Fixed Output	Minimize the Present Worth of costs or other inputs.
Neither Input nor Output Fixed	Maximize [Present Worth of benefits *minus* Present Worth of costs] or, stated another way: Maximize Net Present Worth.

Application of Present Worth

Present worth analysis is most frequently used to determine the present value of future money receipts and disbursements. We might want to know, for example, the present worth of an income producing property, like an oil well. This should provide an estimate of the price at which the property could be bought or sold.

An important restriction in the use of present worth calculations is that there must be a common analysis period when comparing alternatives. It would be incorrect, for example, to compare the present worth (PW) of cost of Pump *A*, expected to last 6 years, with the PW of cost of Pump *B*, expected to last 12 years.

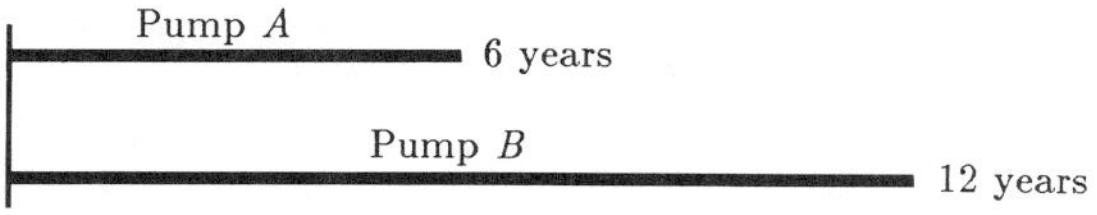

Improper Present Worth Comparison

In situations like this, the solution is either to use some other analysis technique* or to restructure the problem so there is a common analysis period. In the example above, a customary assumption would be that a pump is needed for 12 years and that Pump *A* will be replaced by an identical Pump *A* at the end of 6 years. This gives a 12-year common analysis period.

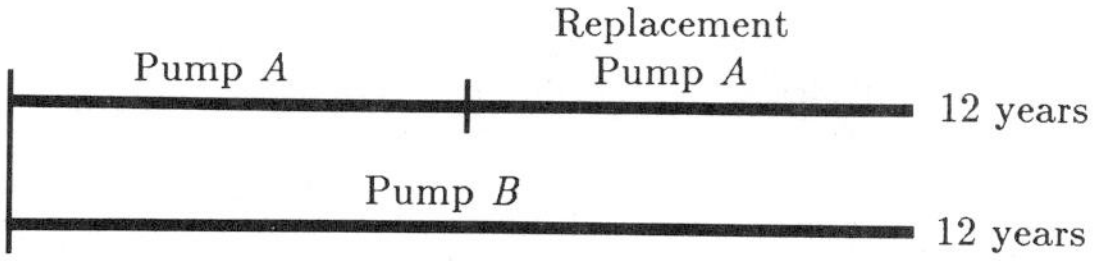

Correct Present Worth Comparison

This approach is easy to use when the different lives of the alternatives have a practical least common multiple life. When this is not true (for example, life of *J* equals 7 years and the life of *K* equals 11 years), some assumptions must be made to select a suitable common analysis period, or the present worth method should not be used.

EXAMPLE 13

Machine *X* has an initial cost of $10,000, annual maintenance of $500 per year, and no salvage value at the end of its four-year useful life. Machine *Y* costs $20,000. The first year there is no maintenance cost. The second year, maintenance is $100, and increases $100 per year in subsequent years. The machine has an anticipated $5000 salvage value at the end of its 12-year useful life.

If interest is 8%, which machine should be selected?

Solution

The analysis period is not stated in the problem. Therefore we select the least common multiple of the lives, or 12 years, as the analysis period.

*Generally the annual cost method is suitable in these situations.

Present Worth of Cost of 12 years of Machine X

$$
\begin{aligned}
&= 10{,}000 + 10{,}000(P/F,8\%,4) + 10{,}000(P/F,8\%,8) + 500(P/A,8\%,12)\\
&= 10{,}000 + 10{,}000(0.7350) + 10{,}000(0.5403) + 500(7.536)\\
&= \$26{,}521
\end{aligned}
$$

Present Worth of Cost of 12 years of Machine Y

$$
\begin{aligned}
&= 20{,}000 + 100(P/G,8\%,12) - 5000(P/F,8\%,12)\\
&= 20{,}000 + 100(34.634) - 5000(0.3971)\\
&= \$21{,}478
\end{aligned}
$$

Choose Machine Y with its smaller PW of Cost.

EXAMPLE 14

Two alternatives have the following cash flows:

	Alternative	
Year	*A*	*B*
0	−$2000	−$2800
1	+800	+1100
2	+800	+1100
3	+800	+1100

At a 5% interest rate, which alternative should be selected?

Solution

Solving by Present Worth analysis:

Net Present Worth (NPW) = PW of benefits − PW of cost

$$
\begin{aligned}
\text{NPW}_A &= 800(P/A,5\%,3) - 2000\\
&= 800(2.723) - 2000\\
&= +178.40
\end{aligned}
$$

$$
\begin{aligned}
\text{NPW}_B &= 1100(P/A,5\%,3) - 2800\\
&= 1100(2.723) - 2800\\
&= +195.30
\end{aligned}
$$

To maximize NPW, choose Alternative B.

Capitalized Cost

In the special situation where the analysis period is infinite ($n = \infty$), an analysis of the present worth of cost is called *capitalized cost.* There are a few public projects where the analysis period is infinity. Other examples would be permanent endowments and cemetery perpetual care.

When n equals infinity, a present sum P will accrue interest of Pi for every future interest period. For the principal sum P to continue undiminished (an essential requirement for n equal to infinity), the end-of-period sum A that can be disbursed is Pi.

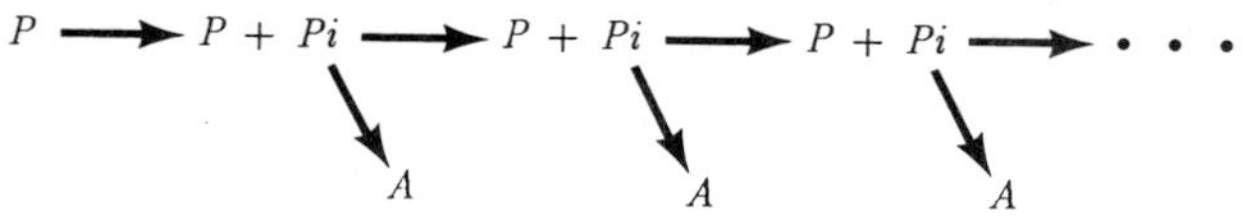

When $n = \infty$, the fundamental relationship between P, A, and i is:

$$A = Pi$$

Some form of this equation is used whenever there is a problem with an infinite analysis period.

EXAMPLE 15
In his will, a man wishes to establish a perpetual trust to provide for the maintenance of a small local park. If the annual maintenance is \$7500 per year and the trust account can earn 5% interest, how much money must be set aside in the trust?

Solution
When $n = \infty$, $A = Pi$ or $P = \frac{A}{i}$

$$\text{Capitalized cost } P = \frac{A}{i} = \frac{\$7500}{0.05} = \$150{,}000$$

FUTURE WORTH

In present worth analysis, the comparison is made in terms of the equivalent present costs and benefits. But the analysis need not be made at the present time, it could be made at any point in time: past, present, or future. Although the numerical calculations may look different, the decision is unaffected by the point in time selected. Of course, there are situations where we do want to know what the future situation will be if we take some particular course of action now. When an analysis is made based on some future point in time, it is called future worth analysis.

Category	*Future Worth Criterion*
Fixed Input	Maximize the Future Worth of benefits or other outputs.
Fixed Output	Minimize the Future Worth of costs or other inputs.
Neither Input nor Output Fixed	Maximize [Future Worth of benefits *minus* Future Worth of costs] or, stated another way: Maximize Net Future Worth.

EXAMPLE 16
Two alternatives have the following cash flows:

	Alternative	
Year	*A*	*B*
0	−\$2000	−\$2800
1	+800	+1100
2	+800	+1100
3	+800	+1100

At a 5% interest rate, which alternative should be selected?

Solution
In Example 14, this problem was solved by Present Worth analysis at Year 0. Here it will be solved by Future Worth analysis at the end of Year 3.

Net Future Worth (NFW) = FW of benefits − FW of cost

$$\begin{aligned} NFW_A &= 800(F/A,5\%,3) - 2000(F/P,5\%,3) \\ &= 800(3.152) - 2000(1.158) \\ &= +205.60 \end{aligned}$$

$$\begin{aligned} NFW_B &= 1100(F/A,5\%,3) - 2800(F/P,5\%,3) \\ &= 1100(3.152) - 2800(1.158) \\ &= +224.80 \end{aligned}$$

To maximize NFW, choose Alternative *B*.

ANNUAL COST

The annual cost method is more accurately described as the method of Equivalent Uniform Annual Cost (EUAC) or, where the computation is of benefits, the method of Equivalent Uniform Annual Benefits (EUAB).

Criteria

For each of the three possible categories of problems, there is an annual cost criterion for economic efficiency.

Category	*Annual Cost Criterion*
Fixed Input	Maximize the Equivalent Uniform Annual Benefits. That is, maximize EUAB.
Fixed Output	Minimize the Equivalent Uniform Annual Cost. That is, minimize EUAC.
Neither Input nor Output Fixed	Maximize [EUAB − EUAC].

Application of Annual Cost Analysis

In the section on present worth, we pointed out that the present worth method requires that there be a common analysis period for all alternatives. This same restriction does not apply in all annual cost calculations, but it is important to understand the circumstances that justify comparing alternatives with different service lives.

Frequently an analysis is to provide for a more or less continuing requirement. One might need to pump water from a well, for example, as a continuing requirement. Regardless of whether the pump has a useful service life of 6 years or 12 years, we would select the one whose annual cost is a minimum. And this would still be the case if the pump useful lives were the more troublesome 7 and 11 years, respectively. Thus, if we can assume a continuing need for an item, an annual cost comparison among alternatives of differing service lives is valid.

The underlying assumption made in these situations is that when the shorter-lived alternative has reached the end of its useful life, it can be replaced with an identical item with identical costs, and so forth. This means the EUAC of the initial alternative is equal to the EUAC for the continuing series of replacements.

If, on the other hand, there is a specific requirement in some situation to pump water for 10 years, then each pump must be evaluated to see what costs will be incurred during the analysis period and what salvage value, if any, may be recovered at the end of the analysis period. The annual cost comparison needs to consider the actual circumstances of the situation.

Examination problems are often readily solved by the annual cost method. And the underlying "continuing requirement" is often present, so that an annual cost comparison of unequal-lived alternatives is an appropriate method of analysis.

EXAMPLE 17

Consider the following alternatives:

	A	*B*
First cost	\$5000	\$10,000
Annual maintenance	500	200
End-of-useful-life salvage value	600	1000
Useful life	5 years	15 years

Based on an 8% interest rate, which alternative should be selected?

Solution

Assuming both alternatives perform the same task and there is a continuing requirement, the goal is to minimize EUAC.

Alternative A:

$$\text{EUAC} = 5000(A/P,8\%,5) + 500 - 600(A/F,8\%,5)$$

$$= 5000(0.2505) + 500 - 600(0.1705) = \$1650$$

Alternative B:

$$\text{EUAC} = 10{,}000(A/P,8\%,15) + 200 - 1000(A/F,8\%,15)$$
$$= 10{,}000(0.1168) + 200 - 1000(0.0368) = \$1331$$

To minimize EUAC, select Alternative *B*.

RATE OF RETURN

A typical situation is a cash flow representing the costs and benefits. The rate of return may be defined as the interest rate where

PW of cost = PW of benefits,

EUAC = EUAB,

or PW of cost − PW of benefits = 0.

EXAMPLE 18

Compute the rate of return for the investment represented by the following cash flow:

Year	*Cash Flow*
0	−\$595
1	+250
2	+200
3	+150
4	+100
5	+50

Solution

This declining arithmetic gradient series may be separated into two cash flows for which compound interest factors are available:

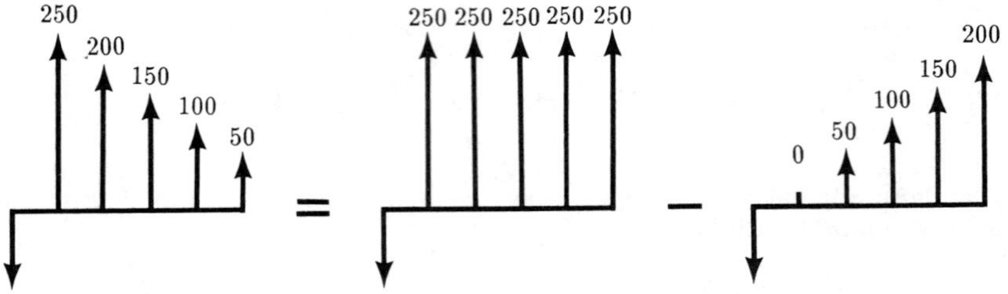

Note that the gradient series factors are based on an *increasing* gradient. Here, the declining cash flow is solved by subtracting an increasing arithmetic gradient, as indicated by the diagram.

PW of cost − PW of benefits = 0

$595 - [250(P/A,i,5) - 50(P/G,i,5)] = 0$

Try $i = 10\%$:

$595 - [250(3.791) - 50(6.862)] = -9.65$

Try $i = 12\%$:

$595 - [250(3.605) - 50(6.397)] = +13.60$

The rate of return is between 10% and 12%. It may be computed more accurately by linear interpolation:

$$\text{Rate of return} = 10\% + (2\%)\left(\frac{9.65 - 0}{13.60 + 9.65}\right) = 10.83\%$$

Rate of Return Criterion for Two Alternatives

Compute the incremental rate of return on the cash flow representing the difference between the two alternatives. Since we want to look at increments of *investment*, the cash flow for the difference between the alternatives is computed by taking the higher initial-cost alternative *minus* the lower initial-cost alternative. If the incremental rate of return is greater than or equal to the predetermined minimum attractive rate of return (MARR), choose the higher-cost alternative; otherwise, choose the lower-cost alternative.

EXAMPLE 19

Two alternatives have the following cash flows:

	Alternative	
Year	*A*	*B*
0	−$2000	−$2800
1	+800	+1100
2	+800	+1100
3	+800	+1100

If 5% is considered the minimum attractive rate of return (MARR), which alternative should be selected?

Solution

These two alternatives were previously examined in Examples 14 and 16 by present worth and future worth analysis. This time, the alternatives will be resolved using rate of return analysis.

Note that the problem statement specifies a 5% minimum attractive rate of raturn (MARR), while Examples 14 and 16 referred to a 5% interest rate. These are really two different ways of saying the same thing: the minimum acceptable time value of money is 5%.

First, tabulate the cash flow that represents the increment of investment between the alternatives. This is done by taking the higher initial-cost alternative minus the lower initial-cost alternative:

Year	Alternative A	Alternative B	Difference between Alternatives B − A
0	−\$2000	−\$2800	−\$800
1	+800	+1100	+300
2	+800	+1100	+300
3	+800	+1100	+300

Then compute the rate of return on the increment of investment represented by the difference between the alternatives:

$$\text{PW of cost} = \text{PW of benefits}$$
$$800 = 300(P/A,i,3)$$
$$(P/A,i,3) = \frac{800}{300} = 2.67$$
$$i \approx 6.1\%$$

Since the incremental rate of return exceeds the 5% MARR, the increment of investment is desirable. Choose the higher-cost Alternative *B*.

Before leaving this example problem, one should note something that relates to the rates of return on Alternative *A* and on Alternative *B*. These rates of return, if computed, are:

	Rate of Return
Alternative *A*	9.7%
Alternative *B*	8.7%

The correct answer to this problem has been shown to be Alternative *B*, and this is true even though Alternative *A* has a higher rate of return. The higher-cost alternative may be thought of as the lower-cost alternative, plus the increment of investment between them. Looked at this way, the higher-cost Alternative *B* is equal to the desirable lower-cost Alternative *A* plus the desirable differences between the alternatives.

The important conclusion is that computing the rate of return for each alternative does not provide the basis for choosing between alternatives. Instead, incremental analysis is required.

EXAMPLE 20
Consider the following:

Year	Alternative A	Alternative B
0	−\$200.0	−\$131.0
1	+77.6	+48.1
2	+77.6	+48.1
3	+77.6	+48.1

If the minimum attractive rate of return (MARR) is 10%, which alternative should be selected?

Solution

To examine the increment of investment between the alternatives, we will examine the higher initial-cost alternative minus the lower initial-cost alternative, or $A - B$.

	Alternative		*Increment*
Year	*A*	*B*	*A − B*
0	−\$200.0	−\$131.0	−\$69.0
1	+77.6	+48.1	+29.5
2	+77.6	+48.1	+29.5
3	+77.6	+48.1	+29.5

Solve for the incremental rate of return:

$$\text{PW of cost} = \text{PW of benefits}$$
$$69.0 = 29.5(P/A,i,3)$$
$$(P/A,i,3) = \frac{69.0}{29.5} = 2.339$$

From Compound Interest Tables, the incremental rate of return is between 12% and 15%. This is a desirable increment of investment hence we select the higher initial-cost Alternative A.

Rate of Return Criterion for Three or More Alternatives

When there are three or more mutually exclusive alternatives, one must proceed following the same general logic presented for two alternatives. The components of incremental analysis are:

1. Compute the rate of return for each alternative. Reject any alternative where the rate of return is less than the given MARR. (This step is not essential, but helps to immediately identify unacceptable alternatives.)

2. Rank the remaining alternatives in their order of increasing initial cost.

3. Examine the increment of investment between the two lowest-cost alternatives as described for the two-alternative problem. Select the best of the two alternatives and reject the other one.

4. Take the preferred alternative from Step 3. Consider the next higher initial-cost alternative and proceed with another two-alternative comparison.

5. Continue until all alternatives have been examined and the best of the multiple alternatives has been identified.

EXAMPLE 21

Consider the following:

	Alternative	
Year	*A*	*B*
0	−$200.0	−$131.0
1	+77.6	+48.1
2	+77.6	+48.1
3	+77.6	+48.1

If the minimum attractive rate of return (MARR) is 10%, which alternative, if any, should be selected?

Solution

One should carefully note that this is a *three*-alternative problem where the alternatives are *A*, *B*, and "Do Nothing."

In this solution we will skip Step 1. Reorganize the problem by placing the alternatives in order of increasing initial cost:

	Do	Alternative	
Year	*Nothing*	*B*	*A*
0	0	−$131.0	−$200.0
1	0	+48.1	+77.6
2	0	+48.1	+77.6
3	0	+48.1	+77.6

Examine the "*B* − Do Nothing" increment of investment:

Year	*B* − Do Nothing	
0	−$131.0 − 0 =	−$131.0
1	+48.1 − 0 =	+48.1
2	+48.1 − 0 =	+48.1
3	+48.1 − 0 =	+48.1

Solve for the incremental rate of return:

$$\text{PW of cost} = \text{PW of benefits}$$
$$131.0 = 48.1(P/A,i,3)$$
$$(P/A,i,3) = \frac{131.0}{48.1} = 2.723$$

From Compound Interest Tables, the incremental rate of return = 5%. Since the incremental rate of return is less than 10%, the *B* − Do Nothing increment is not desirable. Reject Alternative *B*.

Next, consider the increment of investment between the two remaining alternatives:

Year	*A* − Do Nothing	
0	−$200.0 − 0 =	−$200.0
1	+77.6 − 0 =	+77.6
2	+77.6 − 0 =	+77.6
3	+77.6 − 0 =	+77.6

Solve for the incremental rate of return:

$$\begin{aligned} \text{PW of cost} &= \text{PW of benefits} \\ 200.0 &= 77.6(P/A,i,3) \\ (P/A,i,3) &= \frac{200.0}{77.6} = 2.577 \end{aligned}$$

The incremental rate of return is 8%. Since the rate of return on the A − Do Nothing increment of investment is less than the desired 10%, reject the increment by rejecting Alternative A. We select the remaining alternative: Do nothing!

If you have not already done so, you should go back to Example 20 and see how the slightly changed wording of the problem radically altered it. Example 20 required the choice between two undesirable alternatives. Example 21 adds the Do-nothing alternative which is superior to A or B.

EXAMPLE 22

Consider four mutually exclusive alternatives:

	Alternative			
	A	*B*	*C*	*D*
Initial Cost	$400.0	$100.0	$200.0	$500.0
Uniform Annual Benefit	100.9	27.7	46.2	125.2

Each alternative has a five-year useful life and no salvage value. If the minimum attractive rate of return (MARR) is 6%, which alternative should be selected?

Solution

Mutually exclusive is where selecting one alternative precludes selecting any of the other alternatives. This is the typical "textbook" situation. The solution will follow the several steps in incremental analysis.

1. The rate of return is computed for the four alternatives.

Alternative	*A*	*B*	*C*	*D*
Computed rate of return	8.3%	11.9%	5%	8%

Since Alternative C has a rate of return less than the MARR, it may be eliminated from further consideration.

2. Rank the remaining alternatives in order of increasing intial cost and examine the increment between the two lowest cost alternatives.

Alternative	*B*	*A*	*D*
Initial Cost	$100.0	$400.0	$500.0
Uniform Annual Benefit	27.7	100.9	125.2

	$A - B$
Δ Initial Cost	$300.0
Δ Uniform Annual Benefit	73.2
Computed Δ rate of return	7%

Since the incremental rate of return exceeds the 6% MARR, the increment of investment is desirable. Alternative A is the better alternative.

3. Take the preferred alternative from the previous step and consider the next higher-cost alternative. Do another two-alternative comparison.

	$D - A$
Δ Initial Cost	\$100.0
Δ Uniform Annual Benefit	24.3
Computed Δ rate of return	6.9%

The incremental rate of return exceeds MARR, hence the increment is desirable. Alternative D is preferred over Alternative A.

Conclusion: Select Alternative D. Note that once again the alternative with the highest rate of return (Alt. B) is *not* the proper choice.

BENEFIT-COST RATIO

Generally in public works and governmental economic analyses, the dominant method of analysis is called benefit–cost ratio. It is simply the ratio of benefits divided by costs, taking into account the time value of money.

$$\text{B/C} = \frac{\text{PW of benefits}}{\text{PW of cost}} = \frac{\text{Equivalent Uniform Annual Benefits}}{\text{Equivalent Uniform Annual Cost}}$$

For a given interest rate, a B/C ratio ≥ 1 reflects an acceptable project. The method of analysis using B/C ratio is parallel to that of rate of return analysis. The same kind of incremental analysis is required.

B/C Ratio Criterion for Two Alternatives

Compute the incremental B/C ratio for the cash flow representing the increment of investment between the higher initial-cost alternative and the lower initial-cost alternative. If this incremental B/C ratio is ≥ 1, choose the higher-cost alternative; otherwise, choose the lower-cost alternative.

B/C Ratio Criterion for Three or More Alternatives

Follow the logic for rate of return, except that the test is whether or not the incremental B/C ratio is ≥ 1.

EXAMPLE 23

Solve Example 22 using Benefit–Cost ratio analysis. Consider four mutually exclusive alternatives:

		Alternative		
	A	*B*	*C*	*D*
Initial Cost	$400.0	$100.0	$200.0	$500.0
Uniform Annual Benefit	100.9	27.7	46.2	125.2

Each alternative has a five-year useful life and no salvage value. Based on a 6% interest rate, which alternative should be selected?

Solution

1. B/C ratio computed for the alternatives:

Alt. A $$\text{B/C} = \frac{\text{PW of benefits}}{\text{PW of cost}} = \frac{100.9(P/A,6\%,5)}{400} = 1.06$$

B $$\text{B/C} = \frac{27.7(P/A,6\%,5)}{100} = 1.17$$

C $$\text{B/C} = \frac{46.2(P/A,6\%,5)}{200} = 0.97$$

D $$\text{B/C} = \frac{125.2(P/A,6\%,5)}{500} = 1.05$$

Alternative C with a B/C ratio less than 1 is eliminated.

2. Rank the remaining alternatives in order of increasing intial cost and examine the increment of investment between the two lowest cost alternatives.

Alternative	*B*	*A*	*D*
Initial Cost	$100.0	$400.0	$500.0
Uniform Annual Benefit	27.7	100.9	125.2

	$A - B$
Initial Cost	$300.0
Uniform Annual Benefit	73.2

$$\text{Incremental B/C ratio} = \frac{73.2(P/A,6\%,5)}{300} = 1.03$$

The incremental B/C ratio exceeds 1.0 hence the increment is desirable. Alternative A is preferred over B.

3. Do the next two-alternative comparison.

	Alternative		Increment
	A	*D*	$D - A$
Initial Cost	$400.0	$500.0	$100.0
Uniform Annual Benefit	100.9	125.2	24.3

$$\text{Incremental B/C ratio} = \frac{24.3(P/A,6\%,5)}{100} = 1.02$$

The incremental B/C ratio exceeds 1.0, hence Alternative D is preferred.

Conclusion: Select Alternative D.

BREAKEVEN ANALYSIS

In business, "breakeven" is defined as the point where income just covers the associated costs. In engineering economics, the breakeven point is more precisely defined as the point where two alternatives are equivalent.

EXAMPLE 24

A city is considering a new \$50,000 snowplow. The new machine will operate at a savings of \$600 per day, compared to the equipment presently being used. Assume the minimum attractive rate of return (interest rate) is 12% and the machine's life is 10 years with zero resale value at that time. How many days per year must the machine be used to make the investment economical?

Solution

This breakeven problem may be readily solved by annual cost computations. We will set the equivalent uniform annual cost of the snowplow equal to its annual benefit, and solve for the required annual utilization.

Let X = breakeven point = days of operation per year.

$$\text{EUAC} = \text{EUAB}$$
$$50{,}000(A/P,12\%,10) = 600X$$
$$X = \frac{50{,}000(0.1770)}{600} = 14.7 \text{ days/year}$$

DEPRECIATION

Depreciation of capital equipment is an important component of many after-tax economic analyses. For this reason, one must understand the fundamentals of depreciation accounting.

Depreciation is defined, in its accounting sense, as the systematic allocation of the cost of a capital asset over its useful life. *Book value* is defined as the original cost of an asset, minus the accumulated depreciation of the asset.

In computing a schedule of depreciation charges, three items are considered:

1. Cost of the property, P;
2. Depreciable life in years, n;
3. Salvage value of the property at the end of its depreciable life, S.

Straight Line Depreciation

$$\text{Depreciation charge in any year} = \frac{P - S}{n}$$

Sum-Of-Years-Digits Depreciation

$$\begin{array}{c}\text{Depreciation charge}\\ \text{in any year}\end{array} = \frac{\text{Remaining Depreciable Life at Beginning of Year}}{\text{Sum of Years Digits for Total Useful Life}}(P - S)$$

$$\text{where Sum Of Years Digits} = 1 + 2 + 3 + \cdots + n = \frac{n}{2}(n + 1)$$

Double Declining Balance Depreciation

$$\text{Depreciation charge in any year} = \frac{2}{n}(P - \text{Depreciation charges to date})$$

Accumulated Cost Recovery System (ACRS) Depreciation

ACRS depreciation is based on a property class life which is generally less than the actual useful life of the property and on zero salvage value. The varying depreciation percentage to use must be read from a table (based on declining balance with conversion to straight line). Unless one knows the proper ACRS property class, and has an ACRS depreciation table, the depreciation charge in any year cannot be computed.

EXAMPLE 25

A piece of machinery costs $5000 and has an anticipated $1000 salvage value at the end of its five-year depreciable life. Compute the depreciation schedule for the machinery by:

(a) Straight line depreciation;

(b) Sum-of-years-digits depreciation;

(c) Double declining balance depreciation.

Solution

$$\text{Straight line depreciation} = \frac{P - S}{n} = \frac{5000 - 1000}{5} = \$800$$

Sum-of-years-digits depreciation:

$$\text{Sum-of-years-digits} = \frac{n}{2}(n + 1) = \frac{5}{2}(6) = 15$$

$$\text{1st year depreciation} = \frac{5}{15}(5000 - 1000) = \quad \$1333$$

$$\text{2nd year depreciation} = \tfrac{4}{15}(5000 - 1000) = 1067$$

$$\text{3rd year depreciation} = \tfrac{3}{15}(5000 - 1000) = 800$$

$$\text{4th year depreciation} = \tfrac{2}{15}(5000 - 1000) = 533$$

$$\text{5th year depreciation} = \tfrac{1}{15}(5000 - 1000) = \underline{267}$$

$$\$4000$$

Double declining balance depreciation:

$$\text{1st year depreciation} = \tfrac{2}{5}(5000 - 0) = \$2000$$

$$\text{2nd year depreciation} = \tfrac{2}{5}(5000 - 2000) = 1200$$

$$\text{3rd year depreciation} = \tfrac{2}{5}(5000 - 3200) = 720$$

$$\text{4th year depreciation} = \tfrac{2}{5}(5000 - 3920) = 432$$

$$\text{5th year depreciation} = \tfrac{2}{5}(5000 - 4352) = \underline{259}$$

$$\$4611$$

Since the problem specifies a $1000 salvage value, the total depreciation may not exceed $4000. The double declining balance depreciation must be stopped in the 4th year when it totals $4000.

The depreciation schedules computed by the three methods are as follows:

Year	*Straight Line*	*Sum-Of-Years-Digits*	*Double Declining Balance*
1	$800	$1333	$2000
2	800	1067	1200
3	800	800	720
4	800	533	80
5	800	267	0
	$4000	$4000	$4000

INCOME TAXES

Income taxes represent another of the various kinds of disbursements encountered in an economic analysis. The starting point in an after-tax computation is the before-tax cash flow. Generally, the before-tax cash flow contains three types of entries:

1. Disbursements of money to purchase capital assets. These expenditures create no direct tax consequence for they are the exchange of one asset (cash) for another (capital equipment).

2. Periodic receipts and/or disbursements representing operating income and/or expenses. These increase or decrease the year-by-year tax liability of the firm.

3. Receipts of money from the sale of capital assets, usually in the form of a salvage value when the equipment is removed. The tax consequence depends on the relationship between the book value (cost − depreciation taken) of the asset and its salvage value.

Situation	*Tax Consequence*
Salvage value > Book value	Capital gain on difference
Salvage value = Book value	No tax consequence
Salvage value < Book value	Capital loss on difference

After the before-tax cash flow, the next step is to compute the depreciation schedule for any capital assets. Next, taxable income is the taxable component of the before-tax cash flow minus the depreciation. Then, the income tax is the taxable income times the appropriate tax rate. Finally, the after-tax cash flow is the before-tax cash flow adjusted for income taxes.

To organize these data, it is customary to arrange them in the form of a cash flow table, as follows:

Year	*Before-tax cash flow*	*Depreciation*	*Taxable income*	*Income taxes*	*After-tax cash flow*
0	·				·
1	·	·	·	·	·
·	·	·	·	·	·

EXAMPLE 26

A corporation expects to receive \$32,000 each year for 15 years from the sale of a product. There will be an initial investment of \$150,000. Manufacturing and sales expenses will be \$8067 per year. Assume straight line depreciation, a 15-year useful life and no salvage value. Use a 46% income tax rate.

Determine the projected after-tax rate of return.

Solution

$$\text{Straight line depreciation} = \frac{P - S}{n} = \frac{150{,}000 - 0}{15}$$

$$= \$10{,}000 \text{ per year}$$

Year	*Before-tax cash flow*	*Depreciation*	*Taxable income*	*Income taxes*	*After-tax cash flow*
0	−150,000				−150,000
1	+23,933	10,000	13,933	−6,409	+17,524
2	+23,933	10,000	13,933	−6,409	+17,524
·	·	·	·	·	·
·	·	·	·	·	·
·	·	·	·	·	·
15	+23,933	10,000	13,933	−6,409	+17,524

Take the after-tax cash flow and compute the rate of return at which PW of cost equals PW of benefits.

$$150{,}000 = 17{,}524(P/A,i,15)$$
$$(P/A,i,15) = \frac{150{,}000}{17{,}524} = 8.559$$

From Compound Interest Tables, $i = 8\%$.

Professors' Examination Problems and Solutions

1
INTRODUCTORY PROBLEMS
BASIC CONCEPTS

** **1-1**

Many engineers earn high salaries for creating profits for their employers, then find themselves at retirement time insufficiently prepared financially. This may be because in college courses emphasis seldom is placed on using engineering economics for the direct personal benefit of the engineer. Among the goals of every engineer should be assuring that adequate funds will be available for anticipated personal needs at retirement.

A realistic goal of retiring at age 60 with a personal net worth in excess of one million dollars can be accomplished by several methods. A recent independent study ranked the probability of success of the following methods of personal wealth accumulation. Discuss and decide the ranking order of the following five methods.

(a) Purchase as many lottery tickets as possible with money saved from salary.
(b) Place money saved from salary in a bank savings account.
(c) Place all money saved from a salary in a money market account.
(d) Invest saved money into rental properties and spend evenings, weekends and vacations repairing and managing.
(e) Invest all money saved into stock market securities, and study investments 10 to 15 hours per week.

****

Independent studies can be misleading. If Julia McNeese of Roseburg, Oregon were asked to rank wealth accumulation methods, (a) would head her list. Julia recently won one million dollars in a Canadian lottery. A workaholic with handyman talent might select (d) as his Number 1 choice. Lots of people have become millionaires by investing in real estate. The important thing is to learn about the many investment vehicles available and then choose the one or the several most suitable for you.

1-2 **

A food processor is considering the development of a new line of product. Depending on the quality of raw material, he can expect different yields process-wise, and the quality of the final products will also change considerably. The product development department has identified three alternatives, and produced them in a pilot scale. The marketing department has used those samples for surveys to estimate potential sales and pricing strategies. The three alternatives would use exising equipment, but different process conditions and specifications, and they are summarized as follows. Indicate which alternative seems to be the best according to the estimated data, if the objective is to maximize total profit per year.

	Alternative		
	1	2	3
Lbs of raw material A per unit of product	0.05	0.07	0.075
" " " B " "	0.19	0.18	0.26
" " " C " "	0.14	0.12	0.17
Other processing costs ($/unit product)	$0.16	$0.24	$0.23
Expected wholesale price ($/unit product)	0.95	1.05	1.25
Projected volume of sales (units of product)	1,000,000	1,250,000	800,000

Cost of raw material	A	$3.45/lb
" " "	B	1.07
" " "	C	1.88

****

	Alt. 1	Alt. 2	Alt. 3
Cost of Raw Matl. A ($/unit prod)	.05 x 3.45 = .1725	.2415	.2587
Cost of Raw Matl. B ($/unit prod.)	.19 x 1.07 = .2033	.1926	.2782
Cost of Raw Matl. C ($/unit prod.)	.14 x 1.88 = .2632	.2256	.3196
Other processing cost ($/unit prod.)	.16	.24	.23
Total Cost ($/unit prod.)	0.799	.8997	1.0865
Wholesale price ($/unit prod.)	0.95	1.05	1.25
Profit per unit	0.151	0.1503	0.1635
Projected sales (units of prod)	1,000,000	1,250,000	800,000
Projected profits	151,000	187,875	130,800

Therefore, choose alternative 2

** **1-3**

Consider the previous problem. When asked about the precision of the given figures, the marketing department indicated the actual sales results could change plus/minus 10% from the forecast. Similarly, product development indicated the actual production costs may vary 3% from the pilot-based calculations. Is your choice of the best alternative the same as you found in the previous problem?

****

Alternative 1:

Cost

max. = 1.03 x .799 = .82297

min. = 0.97 x .799 = .77503

Profit

min. = .95 − .82297 = .12703

max. = .95 − .77503 = .17497

Sales

max. = 1.1 x 1,000,000 = 1,100,000

min. = 0.9 x 1,000,000 = 900,000

Sales x Profit = TOTAL PROFIT

max. = 1,100,000 x .17497 = $192,467

min. = 900,000 x .12703 = $114,327

Alternative 2:

max. total profit = (1,250,000 x 1.1)(1.05 − .8997 x 0.97) = 243,775

min. total profit = (1,250,000 x 0.9)(1.05 − .8997 x 1.03) = 138,723

Alternative 3:

max. total profit = (800,000 x 1.1)(1.25 − 1.0865 x 0.97) = 172,564

min. total profit = (800,000 x 0.9)(1.25 − 1.0865 x 1.03) = 94,252

Although alternative 2 still gives the largest profits, the data's precision is not good enough to tell for sure that it will actually be the best choice since the max. profits for 1 & 3 are larger than the min. profits of 2.

** **1-4**

Car A initially costs $500 more than Car B, but it consumes 0.04 gallons/mile versus 0.05 gallons/mile for B. Both last 8 years and B's salvage value is $100 smaller than A's. Fuel costs $1.70 per gallon. Other things being equal, beyond how many miles of use per year (= X) does A become preferable to B?

****

$$-500 + 100 + (.05 - .04)(1.70)(8)X = 0$$

$$-400 + 0.136X = 0$$

$$X = \frac{400}{0.136} = \underline{2{,}941 \text{ miles/year}}$$

1-5 **

The following letter was a reply from Benjamin Franklin to Joseph Priestley, a friend of Franklin's. Priestley had been invited to become the librarian for the Earl of Shelburne and had asked for Franklin's advice. What engineering economy principle does Franklin suggest Priestley use to aid in making his decision?

London, September 19, 1772

Dear Sir:

In the affair of so much importance to you wherein you ask my advice, I cannot, for want of sufficient premises, advise you what to determine, but if you please I will tell you how. When these difficult cases occur, they are difficult chiefly because while we have them under consideration, all the reasons Pro and Con are not present to the mind at the same time; but sometimes one set present themselves, and at other times another, the first being out of sight. Hence the various purposes or inclination that alternately prevail, and the uncertainty that perplexes us.

To get over this, my way is to divide a half a sheet of paper by a line into two columns; writing over the one PRO and over the other CON. Then during three or four days' consideration I put down under the different heads short hints of the different motives that at different times occur to me, for or against the measure. When I have thus got them all together in one view, I endeavour to estimate their respective weights; and where I find two (one on each side) that seem equal, I strike them both out. If I find a reason Pro equal to some two reasons Con, I strike out the three. If I judge some two reasons Con equal to three reasons Pro, I strike out the five; and thus proceeding I find at length where the balance lies; and if after a day or two of further consideration, nothing new that is of importance occurs on either side, I come to a determination accordingly. And though the weight of the reasons cannot be taken with the precision of algebraic quantities, yet when each is thus considered separately and comparatively and the whole lies before me, I think I can judge better, and am less likely to make a rash step; and in fact I have found great advantage from this kind of equation in what may be called moral or prudential algebra.

Wishing sincerely that you may determine for the best, I am ever, my dear friend, your most affectionately...

s/Ben Franklin

****

Decisions should be based on the differences between the alternatives. Here the alternatives are taking the job (Pro) and not taking the job (Con).

** **1-6**

Assume that you are employed as an engineer for Wreckall Engineering, Inc., a firm specializing in demolition of high rise buildings. The firm has won a bid to tear down a 30-story building in a heavily developed downtown area. The crane owned by the company only reaches to 29 stories. Your boss asks you to perform an economic analysis of buying a new crane to complete the job. How would you handle the analysis?

****

The important point of this problem is to realize that your boss may not have recognized what the true problem is in this case. To buy a new crane is only one alternative, and quite likely not the best alternative.

Other alternatives:
- extension on current crane
- ramp for current crane
- rent a crane to remove top story
- explosive demolition
- ⋮

If this is a fixed output project (e.g., fixed fee for demolishing building) we want to minimize costs. Weigh alternatives using economic criteria to choose the best alternative.

** **1-7**

The total cost of a building (TC) is given by

$$TC = (200 + 80X + 2X^2)A$$

where X = Number of floors
A = Floor area in ft^2/floor

If the total number of square feet required is 10^6, what is the optimal (minimum cost) number of floors?

****

$$TC = (200 + 80x + 2x^2)(10^6/x)$$

$$dTC/dx = (10^6)(-200/x^2 + 2) = 0$$

$$x^* = \sqrt{200/2} = \sqrt{100} = 10 \text{ floors}$$

1-8 **

By saving and investing, wisely or luckily or both, Helen finds she has accumulated $400,000 in savings while her salaried position is providing her with $40,000 per year, including benefits, and after income taxes and other deductions.

Helen's salaried position is demanding and allows her little free time, but the desire to pursue other interests has become very stong. What would be your advice to her if you were asked?

****

First, Helen should decide what annual income she needs to provide herself with the things she wants. Depending on her age, she might be able to live on the interest income (maybe 10% x $400,000 = $40,000), or a combination of interest and principal. The important thing that Helen should realize is that it may be possible for her to lead a more fulfilling lifestyle if she is fully aware of the time value of money. There are many people with large sums of money in bank checking accounts (drawing no interest) because they can write "free" checks.

1-9 **

Charles belongs to a square dance club that meets twice each month and has quarterly dues of $9.00 per person. The club moved its meeting place to a location with increased cost. To offset the cost each member agrees to pay 50 cents each time they attend the meeting. Later the treasurer suggests that the quarterly dues be increased to $12.00 per person as an alternative to the meeting charge. Discuss the consequences of the proposal. Do you think the club members would agree to the proposal?

****

The members who attend regularly would pay the same amount with the new dues as with the older method of $9.00 plus 50 cents per meeting. Many would like the added advantage of covering their quarterly expenses in one check. The members who attend infrequently would pay more by the new method and might oppose the action.

Since the people who attend infrequently are in the minority in this club, the members voted to approve the proposal.

** **1-10**

Sam decides to buy a cattle ranch and leave the big city rat race. He locates an attractive 500-acre spread in Montana for $1000 per acre, that includes a house, a barn, and other improvements. Sam's studies indicate that he can run 200 cow-calf pairs and be able to market 180 500-pound calves per year. Sam, being rather thorough in his investigation, determines that he will need to purchase an additional $95,000 worth of machinery. He expects that supplemental feeds, medications and veterinary bills will be about $50 per cow per year. Property taxes are $4000 per year, and machinery upkeep and repairs are expected to run $3000 per year.

If interest is 10% and Sam would like a net salary of $10,000 per year, how much will he have to get for each 500-pound calf?

****

Land Cost:	500 Ac. x $1,000/Ac. =	$500,000
Machinery:	lump sum =	95,000
Total Fixed Cost		$595,000

Assume land and machinery to have a <u>very</u> long life

At 10% Annual Cost = (.10)($595,000) = $59,500

Other Annual Costs:

Feeds, medications, vet bills $50 x 200 =	$10,000
Property taxes	4,000
Upkeep & Repairs	3,000
Salary	10,000
Total Annual Cost	$86,500

Net sale price of each calf would have to be:

$$\frac{\$86,500}{180} = \$480.56$$

Note: If Sam were to invest his $595,000 in a suitable investment vehicle yeilding 10% interest his salary would be almost six times greater and he could go fishing instead of punching cows.

1-11 **

Smith borrowed $1000 from Jones for a 5-year period at a yearly interest rate of 14%. Smith promised to make a simple interest payment each 6 months and to repay the loan after 5 years. Draw a cash flow diagram to show Smith's payments.

****

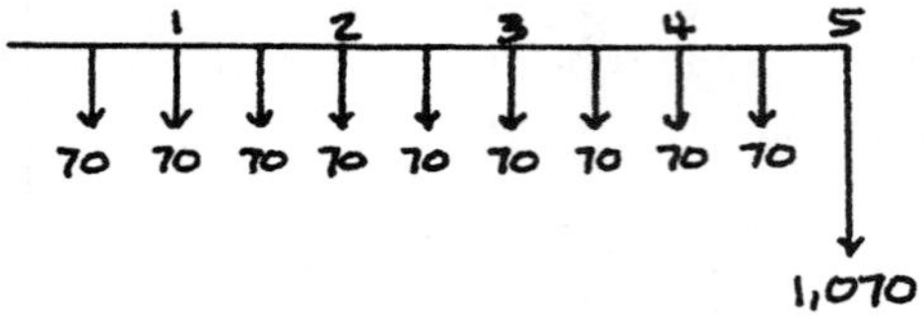

$I = Prt$

$= 1000 \times 0.14 \times 1/2 = \$70$

1-12 **

What is the annual interest rate if a simple interest loan of $15,000 for four months yields $975 interest?

****

$I = Prt$

$\$975 = \$15{,}000 \times r \times 4/12 \text{ yr.}$

$r = 975/15{,}000 \times 12/4 = .195 = \underline{19.5\%/\text{yr}}$

$i = rt$

$= .195 \times 4/12 = .065 = \underline{6.5\% \text{ for 4 months}}$

2 INTEREST PROBLEMS AND EQUIVALENCE

** **2-1**

Solve for the unknown value. Be sure to show your work.

P = 1000 i = 12% n = 5 F = ?

****

$$F = P(F/P, i\%, n) = 1{,}000(F/P, 12\%, 5) = \$1{,}762.00$$

** **2-2**

If you had $1000 now and invested it at 6%, how much would it be worth 12 years from now?

****

$$F = P(F/P, i\%, n) = 1{,}000(F/P, 6\%, 12) = 1{,}000(2.012)$$
$$= \$2{,}012.00$$

** **2-3**

The effective interest rate is 19.56%. If there are 12 compounding periods per year, what is the nominal interest rate?

****

$$i_{eff} = \left(1+\frac{r}{m}\right)^m - 1 \Rightarrow \frac{r}{m} = (1+i_{eff})^{1/m} - 1$$
$$= (1.1956)^{1/12} - 1$$
$$= 1.5\%$$

$$r = 12 \times 1.5 = 18\%$$

2-4 **

Given: Nominal interest = 9%, compounded monthly

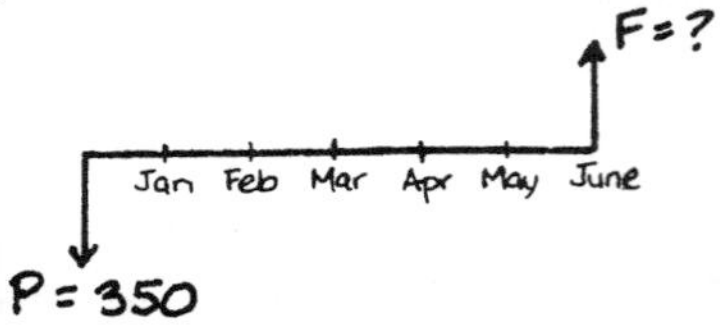

Find: (a) F
(b) i_{eff}

****

Soln: (a) $F = 350(F/P, .75\%, 6) = 366.10$

(b) $i_{eff} = (1+i)^m - 1 = (1.0075)^{12} - 1 = 9.38\%$

2-5 **

A young engineer wishes to buy a house but only can afford monthly payments of $500. Thirty year loans are available at 12% interest compounded monthly. If she can make a $5000 downpayment, what is the price of the most expensive house that she can afford to purchase?

****

$\frac{12\%}{12} = 1\%$ at $(30)(12) = 360$

$P - 5{,}000 = 500(P/A, 1\%, 360) = 500(97.218) = 48{,}609$

$P = \$53{,}609$

2-6 **

Mr. Beach deposited $200,000 in the Lawrence National Bank. If the bank pays 8% interest, how much will he have in the account at the end of 10 years?

****

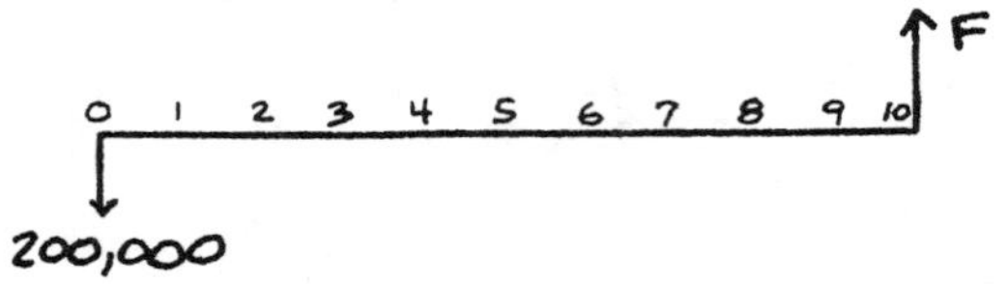

$F = 200{,}000(F/P, 8\%, 10) = 200{,}000(2.159)$

$F = \$431{,}800$

*** **2-7**

Given:

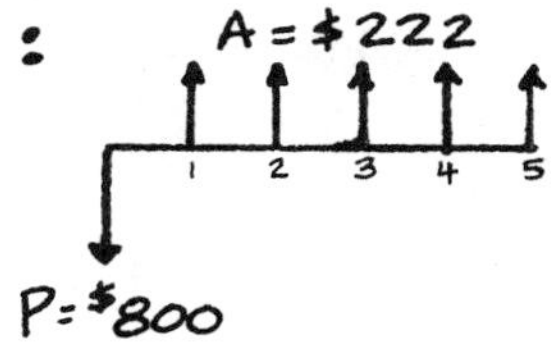

Find: i %

****

Soln: $P = A(P/A, i\%, 5)$

$(P/A, i\%, 5) = \frac{800}{222}$

$= 3.6$

From tables

<u>$i = 12\%$</u>

*** **2-8**

How much should Reggie invest in a fund which will pay 9%, compounded continuously, if he wishes to have $600,000 in the fund at the end of 10 years?

****

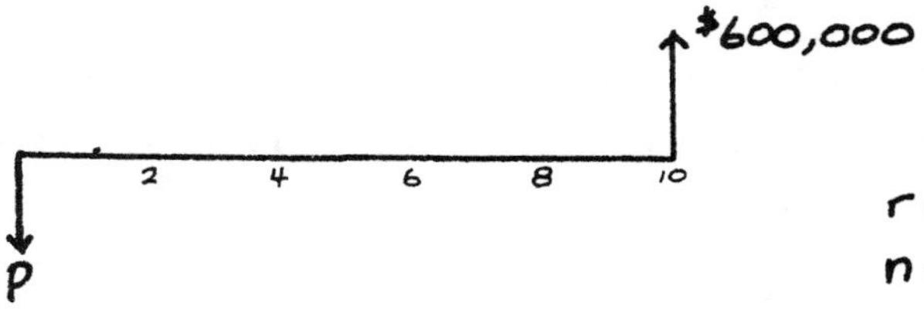

$r = 0.09$

$n = 10$

$P = F(P/F, 9\%, \infty) = Fe^{-rn} = 600{,}000(0.40657)$

<u>$= \$243{,}941.80$</u>

*** **2-9**

Solve for the unknown interest rate.

P = 1000 n = 10 years A = 238.50 i = ?

****

$A = P(A/P, i\%, n) \Rightarrow \frac{A}{P} = (A/P, i\%, 10) = \frac{238.50}{1{,}000} = 0.2385$

from tables; $i = 20\%$

2-10 **

How much will accumulate in an Individual Retirement Account (IRA) in 15 years if $500 is deposited in the account at the end of each quarter during that time? The account earns 12% interest, compounded quarterly. What is the effective interest rate?

****

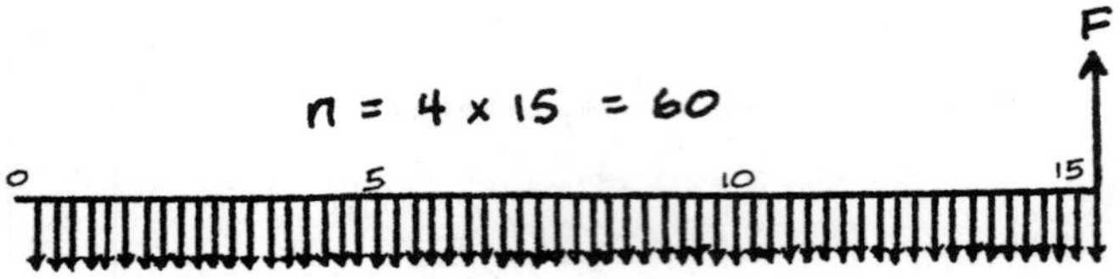

$r = 12\%$

$i = \frac{12}{4} = 3\%$

$F = 500(F/A, 3\%, 60) = 500(163.053) = \$\underline{81{,}526.50}$

Effective interest rate $= (1 + .03)^4 - 1 = 1.1255 - 1$

$= \underline{12.55\%}$

2-11 **

A continuously compounded loan has what nominal interest rate if the effective interest rate is 25%? Select one of the five choices below. (Closed book problem.)

(a) $e^{1.25}$ (b) $e^{0.25}$

(c) $\log_e(1.25)$ (d) $\log_e(0.25)$

(e) Other (specify)________________________

****

$e^r - 1 = .25$

$e^r = 1.25$

$\log_e(e^r) = \log_e(1.25)$

$r = \log_e(1.25)$

$\therefore$ choose c

2-12 **

Solve for the unknown value. Be sure to show your work.

P = 1000 i = 12% n = 5 A = ?

****

$A = P(A/P, i\%, n) = 1{,}000(A/P, 12\%, 5) = \$277.40$

** **2-13**

To offset the cost of buying a $75,000 house, a couple borrowed $12,500 from their parents at 12% nominal interest, compounded monthly. The loan from their parents is to be paid off in five years in equal monthly payments. The couple has saved $11,250. Their total downpayment is therefore $12,500 + 11,250 = $23,750. The balance will be mortgaged at 15% nominal interest, compounded monthly for 30 years.

Find the combined monthly payment that the couple will be making for the first five years.

****

12,500(A/P, 1%, 60) = 12,500(.0103) = $128.75

borrowed from bank : 75,000 - 23,750 = $51,250

51,250 (A/P, 1¼%, 360) = 51,250(.0126) = 645.75

∴ monthly payments are 128.75 + 645.75 = **$774.50**

** **2-14**

Find C if i = 12%

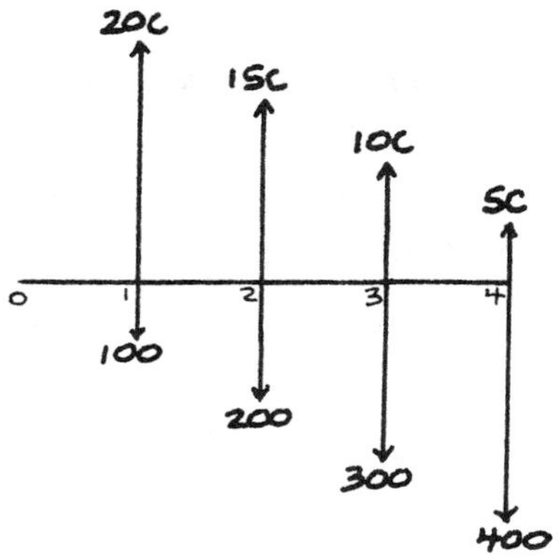

****

20C(P/A, 12%, 4) - 5C(P/G, 12%, 4) = 100(P/A, 12%, 4) + 100(P/G, 12%, 4)

20C(3.037) - 5C(4.127) = 100(3.037) + 100(4.127)

60.74C - 20.635C = 303.7 + 412.7

40.105C = 716.4

C = 17.86

2-15 **

Decide whether each of the three statements below is TRUE or FALSE without referring to your book, notes, or compound interest tables.

(a) If interest is compounded quarterly, the interest period is four months.

(b) (F/A,12%,30) = (F/A,1%,360)

(c) (F/P,i%,10) is greater than (P/F,i%,10) for all values of i% > 0%.

****

(a) FALSE. If interest is compounded quarterly, each interest period is 3 months long.

(b) FALSE. If we assume, for example, we are talking about 30 years or 360 months, the (F/A,12%,30) does not provide monthly compounding of interest. The (F/A,1%, 360) does. (This is a common error among beginning students.)

(c) TRUE. Since i%>0%, (1+i) > 0. Thus (F/P,i%,10) = $(1+i)^{10} > 1$

and (P/F,i%,10) = $(1+i)^{-10} < 1$ for all values of i% > 0%.

2-16 **

A company borrowed $10,000 at 12% interest. The loan was repaid according to the following schedule. Find X, the amount that will pay off the loan at the end of year 5.

Year	Amount
1	$2000
2	2000
3	2000
4	2000
5	X

****

10,000 = 2,000(P/A,12%,4) + x(P/F,12%,5)

10,000 = 2,000(3.037) + x(.5674)

10,000 = 6,074 + x(.5674)

3,926 = x(.5674)

$$x = \frac{3926}{.5674} = \underline{6,919.28}$$

** **2-17**

Decide whether each of the two statements below is TRUE or FALSE without referring to your book, notes, or compound interest tables.

(a) A young engineer calculated that monthly payments of $A are required to pay off a $5000 loan for n years at i% interest, compounded monthly. If the engineer decides to borrow $10,000 instead, her monthly payments will be $2A.

(b) 400(P/A,i%,5) −100(P/G,i%,5) = 400(P/A,i%,4) −100(P/G,i%,4)

****

(a) $A = 5{,}000(A/P, \frac{i\%}{12}, 12n)$

$2A = 10{,}000(A/P, \frac{i\%}{12}, 12n) = 2[5{,}000(A/P, \frac{i\%}{12}, 12n)$

∴ TRUE

(b) For : $400(P/A, i\%, 5) - 100(P/G, i\%, 5)$

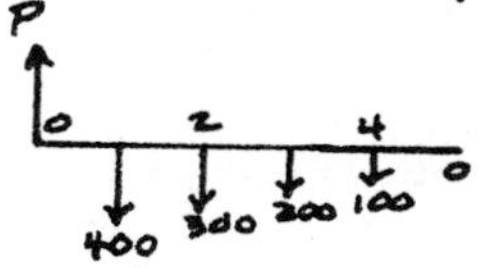

For : $400(P/A, i\%, 4) - 100(P/G, i\%, 4)$

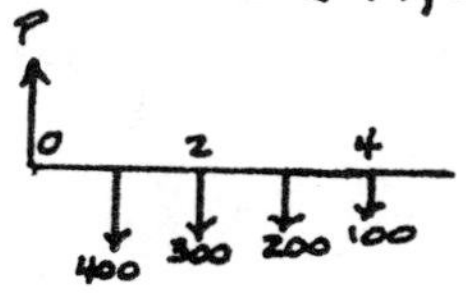

∴ TRUE, since cash flow diagrams are identical

** **2-18**

A continuously compounded loan has what effective interest rate if the nominal interest rate is 25%? Select one of the five choices below. (Closed book problem.)

(a) $e^{1.25}$ (b) $e^{0.25}$

(c) $\log_e(1.25)$ (d) $\log_e(0.25)$

(e) Other (specify)________________

****

$e^{.25} - 1$ ∴ choose e

2-19 **

How much will Reggie accumulate in a bank account if he deposits $3000 at the end of each year for 7 years? Use interest = 10% per annum.

****

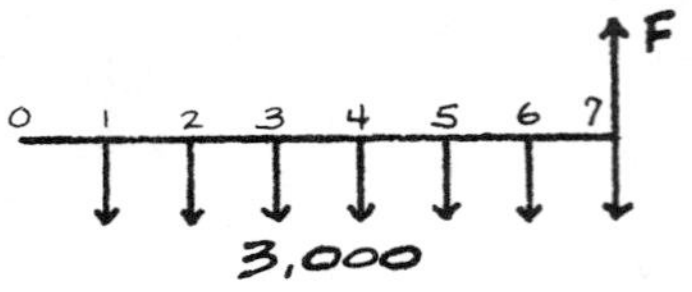

$F = A(F/A, 10\%, 7)$

$= 3{,}000\,(9.487) = \$\underline{28{,}461}.$

2-19A **

You need to borrow $10,000 and the following two alternatives are available at different banks:

(a) Pay $2983 at the end of each year for 5 years, starting at the end of the first year. (5 payments in total.)

(b) Pay $237.90 at the end of each month, for 5 years, starting at the end of the first month. (60 payments in total.)

On the basis of the interest rate being charged in each case, which alternative should you choose?

****

<u>Alternative 1</u>:

$NPW = 0 = 10{,}000 - 2{,}983(P/A, i, 5)$

then $(P/A, i, 5) = \dfrac{10{,}000}{2{,}983} = 3.3523$

from tables; $i \simeq 15\%$ (nominal & effective annual rate since compounded annually.)

<u>Alternative 2</u>:

$n = 5 \times 12 = 60$

$NPW = 0 = 10{,}000 - 237.90(P/A, i, 60)$

then $(P/A, i, 60) = \dfrac{10{,}000}{237.90} = 42.035$

from tables $i = 1.25\%$

the nominal annual interest rate is: $12 \times 1.25 = 15\%$

but the effective interest rate is: $(1 + \frac{0.15}{12})^{12} - 1 = 16.075\%$

Therefore, choose the first alternative.

** **2-20**

Find the Uniform Equivalent for the following cash flow diagram if i = 18%. Use the appropriate gradient and uniform series factors.

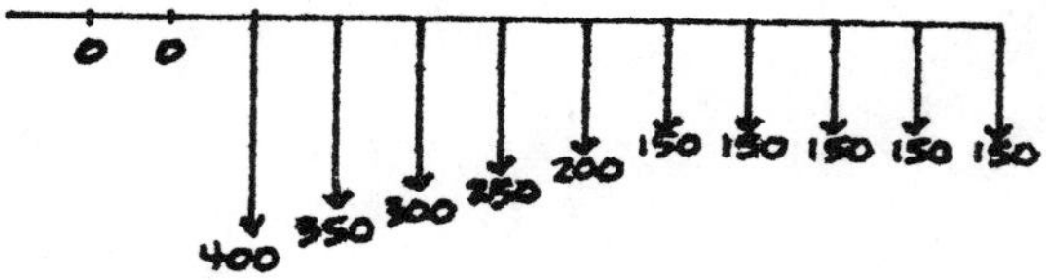

****

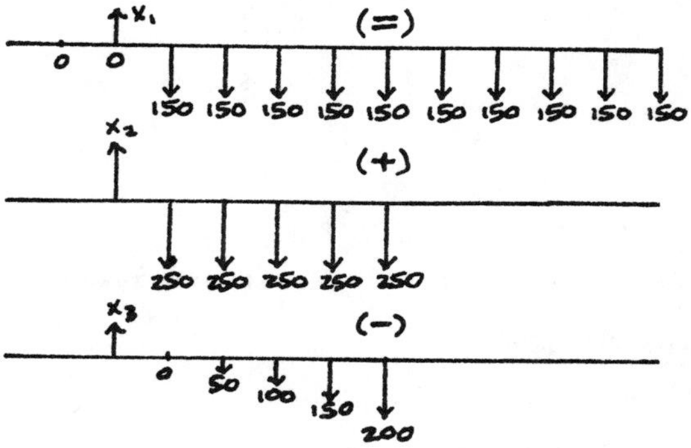

$X_1 = 150(P/A, 18\%, 10) = 150(4.494) = 674.10$

$X_2 = 250(P/A, 18\%, 5) = 250(3.127) = 781.75$

$X_3 = 50(P/G, 18\%, 5) = 50(5.231) = 261.55$

$X = X_1 + X_2 - X_3 = \$1{,}194.30$

$P = 1{,}194.30(P/F, 18\%, 2) = \$1{,}194.30(.7182) = 857.75$

$A = 857.75(A/P, 18\%, 12) = 857.75(.2086) = \underline{\$178.93}$

** **2-20A**

A young woman placed $200 in a savings account paying monthly interest. After one year her balance had grown to $212.64. What was the effective annual interest rate?

****

By definition the effective annual interest rate is determined by the actual interest earned in one year.

$r_e = \dfrac{12.64}{200} = \underline{6.32\%}$

2-21 **

Below is an equation to compute an Equivalent Uniform Cash Flow (EUCF). Determine the values of the net cash flow series that is implied by the equation.

EUCF = [-8000 -8000(P/F,10%,1)](A/P,10%,8)
+2000 +500(P/G,10%,4)(P/F,10%,1)(A/P,10%,8)
+750[(P/F,10%,6) - (P/F,10%,8)](A/P,10%,8)

Time	Net Cash Flow Series
0	
1	
2	
3	
4	
5	
6	
7	
8	

****

t	Net Cash Flow Series				
0	-8000	=	-8000		
1	-6000	=	-8,000 +2,000		
2	2,000	=	+2,000		
3	2,500	=	+2,000	+500	
4	3,000	=	+2,000	+1,000	
5	3,500	=	+2,000	+1,500	
6	2,750	=	+2,000		+750
7	2,000	=	+2,000		
8	1,250	=	+2,000		-750

2-22 **

Given: A situation where the annual interest rate is 5%. When continuous compounding is used, rather than monthly compounding, the nominal interest rate

(Select One)

(a) Increases

(b) Remains the same

(c) Decreases

****

The answer is (b): Remains the same

*** **2-23**

A journalist for a small town newspaper has a weekly column in which he answers questions from the local populace on various financial matters. Below is one such question. Assume you are the journalist and respond to this inquiry. Be brief and specific.

Q: "I put $10,000 in a 12% six month savings certificate. When it matured, I expected to receive interest of $1200 - 12% of $10,000. Instead, I received only $600. Why? Current six month certificates now pay 9% interest. If I put $10,000 in a new six month certificate, I assume (based on my previous experience) I'll get only $450 interest - one half of 9%. Wouldn't my money earn more interest if I deposit it in a savings account paying 5-1/2 percent?"

A:

****

The questioner is confused about nominal vs. effective interest rates. The 9% and 12% rates are nominal annual interest rates compounded semi-annually. The effective semi-annual interest rates are 0.09/2 = 0.045 and 0.12/2 = 0.06 hence the interest earned in 6 months would be 0.045(10,000) = $450 and 0.06(10,000) = $600. The corresponding effective annual interest rates are

$$(1.045)^2 - 1 = 0.092 \text{ and } (1.06)^2 - 1 = 0.1236$$

The interest rate advertised on the savings account typically is also a nominal annual rate. Most accounts pay interest quarterly or continuously, thus the effective annual interest rate would be either

$$(1.01375)^4 - 1 = 0.05614 \quad \text{or} \quad e^{0.055} - 1 = 0.05654$$

neither of which is as good as the six month certificates.

*** **2-24**

Draw a diagram that represents (P/A,i%,4)

****

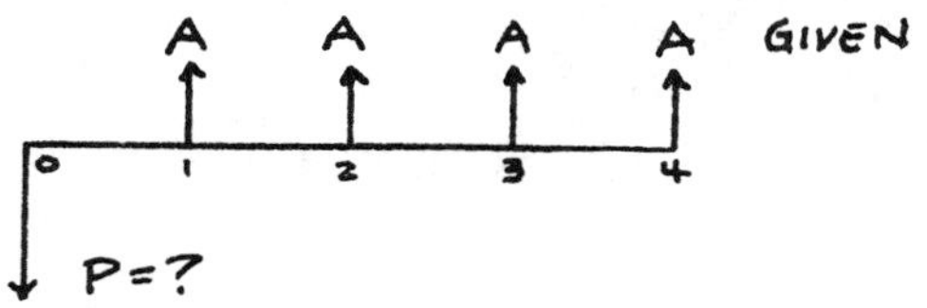

2-25 **

On July 1 and September 1, Joan placed $2000 into an account paying 12% compounded monthly. How much was in the account on October 1?

****

$F = 2{,}000(1+.01)^3 + 2{,}000(1.01)^1 = \underline{\$4080.60}$

(or)

Month	Beg. Bal	Int @ 1%	End Bal	Deposit
July	2,000	20.00	2020.00	0
Aug.	2,020	20.20	2040.20	2000
Sept.	4,040.20	40.40	4080.60	0
Oct.	4,080.60 ←			

2-26 **

A drug dealer will sell dope to his regular customers for $20 immediately or $22 if the payment is deferred one week. What nominal annual interest rate is the dealer receiving?

****

$i = \frac{2}{20} = 10\%$

$r = 52 \times 10\% = \underline{520\%}$

2-27 **

Using a credit card, Ben Spendthrift has just purchased a new stereo system for $975 and will be making payments of $45 per month. If the interest rate is 18% compounded monthly, how long will it take to completely pay off the stereo?

****

$i = \frac{18\%}{12} = 1\frac{1}{2}\%$

$975 = 45(P/A, 1\frac{1}{2}\%, n)$

$(P/A, 1\frac{1}{2}\%, n) = 975/45 = 21.667$

$n = 27$ loan is not completely paid off in 26 months.

There must be a smaller payment in the 27th month.

*** **2-28**

An engineer on the verge of retirement has accumulated savings of $100,000 which are in an account paying 6% compounded quarterly. The engineer wishes to withdraw $6000 each quarter. For how long can she withdraw the full amount?

****

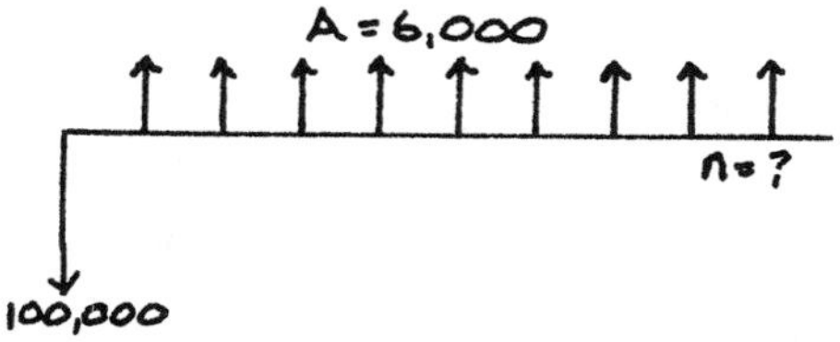

$i = 6\%/4 = 1\tfrac{1}{2}\%$

$A = 6{,}000 = 100{,}000\,(A/P, 1\tfrac{1}{2}\%, n)$

$(A/P, 1\tfrac{1}{2}\%, n) = 0.0600$

From tables <u>n = 19 quarters</u> or $4\tfrac{3}{4}$ years

note: This leaves some money in the account but not enough for a full $6,000 withdrawal

*** **2-29**

Charles puts $25 per month into an account at 9% interest for two years to be used to purchase an automobile. The car he selects then costs more than the amount in the fund. He agrees to pay $50 per month for two more years, at 12% interest, and also makes a cash payment of $283.15. What is the cost of the automobile?

****

$$P = 283.15 + 25\,(F/A, \tfrac{3}{4}\%, 24) + 50\,(P/A, 1\%, 24)$$
$$= 283.15 + 25(26.188) + 50(21.243)$$
$$= 283.15 + 654.70 + 1{,}062.15 = \underline{\$2{,}000 \text{ cost of auto}}$$

*** **2-30**

Explain in one or two sentences why (A/P,i,infinity) = i.

****

In order to have an infinitely long A series, the principal must never be reduced. For this to happen only the interest earned each period may be removed. Removing more than the interest would deplete the principal so that even less interest is available the next period.

2-31 **

Find the value of X which makes the two cash flow series equivalent at 18%.

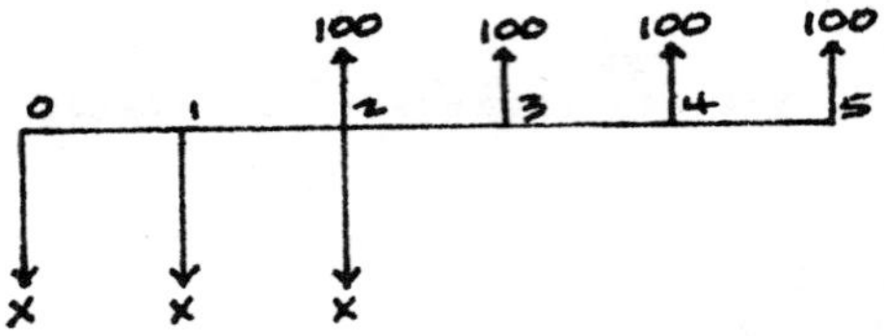

****

Equate the two series at the end of period 2:

$X(F/A, 18\%, 3) = 100 + 100(P/A, 18\%, 3)$

$X(3.572) = 100 + 100(2.174)$

$X = 88.86$

2-32 **

A local bank is advertising that they pay savers 6% compounded monthly, yielding an effective annual rate of 6.168%. If $2000 is placed in savings now and no withdrawals are made, how much interest (to the penny) will be earned in one year?

****

Interest = Effective annual rate x principal

= 0.06168 x 2,000 = $123.36

Monthly compounding is irrelevant when the effective rate is known.

2-33 **

A man decides to put $100 per month into an account paying 12% compounded monthly. Without using formulas or factors (that is, use only basic concepts) determine how much (to the penny) will be in the account immediately after the fourth deposit.

****

Month	Beg. Bal.	Int.@1%	Deposit	End Bal.
1	0	0	100	100
2	100	1.00	100	201.00
3	201.00	2.01	100	303.01
4	303.01	3.03	100	406.04 ←

*** 2-34**

A bank is offering a loan of $20,000 with a nominal interest rate of 12%, payable in 48 months.

(a) Calculate first the monthly payments.

(b) This bank also charges a loan fee of 4% of the amount of the loan, payable at the time of the closing of the loan (that is, at the time they give the money to the borrower). What is the effective interest rate they are charging?

****

a) the monthly payments:

$n = 48$; $i = \frac{12\%}{12} = 1\%$ per period (month)

$A = P(A/P, i, n) = 20{,}000(A/P, 1\%, 48\%)$

$= 20{,}000(.0263) = \underline{526.00}$

b) Actual money received $= P = 20{,}000 - 0.04(20{,}000)$

$= 19{,}200$

But $A = 526.00$; $n = 48$

then $A = P(A/P, i, n)$

$526 = 19{,}200\,(A/P, i, 48)$

$(A/P, i, 48) = \frac{526}{19{,}200} = 0.02739$

for $i = 1\frac{1}{4}\%$ $(A/P, 1\frac{1}{4}\%, 48) = 0.0278$

for $i = 1\%$ $(A/P, 1\%, 48) = 0.0263$

then by interpolation $i \approx 1\% + \frac{1}{4}\%\left(\frac{.0263 - .02739}{.0263 - .0278}\right)$

$i \approx 1.1817\%$

the effective intrest rate $= (1 + 0.011817)^{12} - 1$

$= 0.1514 = \underline{15.14\%}$

*** 2-35**

Find A if A = $3000(A/P,13.5%, infinity).

****

$A = 3{,}000 \times .135 = \$\underline{405}$.

2-36 **

A small company borrowed $10,000 to expand the business. The entire principal of $10,000 will be repaid in 2 years but quarterly interest of $330 must be paid every three months. What nominal annual interest rate is the company paying?

****

Since $330 is interest only for one interest period, then

$i = \frac{330}{10,000} = 3.3\%$ per quarter

$r = 3.3 \times 4 = \underline{13.2\%}$ nominal annual

2-37 **

A store policy is to charge 1-1/4% interest each month on the unpaid balance.

(a) What is the nominal interest?
(b) What is the effective interest?

****

a) $N_i = mi = 12(1.25) = \underline{15\%}$

b) $E_i = (1+i)^n - 1 = (1.0125)^{12} - 1 = \underline{16.075\%}$

2-38 **

Compute the value of Q in the figure below if i = 18%.

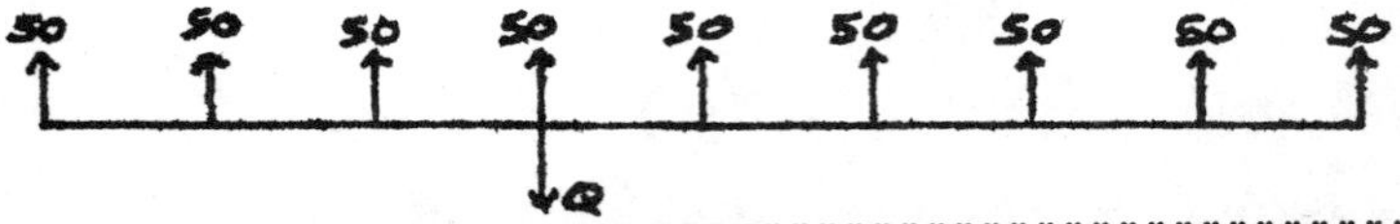

****

$Q = 50(F/A, 18\%, 4) + 50(P/A, 18\%, 5) = 50(5.215) + 50(3.127)$

$= \underline{\$417.10}$

2-39 **

Under what circumstances are the nominal and effective annual interest rates exactly equal; or is this never true?

****

The nominal interest rate equals the effective interest rate when there is yearly (annual) compounding (i.e., m = 1).

** **2-40**

A small company borrowed $10,000 to expand the business. The entire principal of $10,000 will be repaid at the end of two years but quarterly interest of $335 must be paid every three months. What nominal annual interest rate is the company paying?

****

$i = \frac{335}{10{,}000} = 3.35\%$

$r = im = 4 \times 3.35 = 13.40\%$

** **2-41**

E.Z. Marc received a loan of $50 from the S.H. Ark Loan Company which he had to repay one month later with a single payment of $60. What was the nominal annual interest rate for this loan?

****

Interest = \$10 in one month

$i = 10/50 = 20\%$

$r = im = 20 \times 12 = \underline{240\%}$

** **2-42**

A local college parking enforcement bureau issues parking tickets which must be paid within one week. The person receiving the ticket may pay either $5 immediately, or $7 if payment is deferred one week. What nominal interest rate is implied in the arrangement?

****

$i = \frac{7-5}{5} = 40\%$ per week

$r = mi = 52(40) = 2080\%$

** **2-43**

A deposit of $300 was made one year ago into an account paying monthly interest. If the account now has $320.52, what was the effective annual interest rate? Give answer to 1/100 of a percent.

****

$r_e = \frac{20.52}{300} = \underline{6.84\%}$

2-44 **

Henry Fuller purchases a used automobile for $4500. He wishes to limit his monthly payment to $100 for a period of two years. What downpayment must he make to complete the purchase if the interest rate is 15% on the loan?

****

$P = P' + A(P/A, 1\frac{1}{4}\%, 24)$

$4{,}500 = P' + 100(20.624)$

$P' = 4{,}500 - 2{,}062.40 = \$\underline{2{,}437.60 \text{ downpayment}}$

2-45 **

Which is the better investment - a fund that pays 15% compounded annually, or one that pays 14% compounded continuously?

****

$i = 15\%$; $n = 1$ $F = P(1+.15)^1 = 1.1500P$

$r = 14\%$ cont; $n = 1$ $F = Pe^{.14} = 1.1503P$

14% compounded continuously is slightly better

2-46 **

For a nominal interest of 16 percent, what would the effective interest be, if interest is

(a) compounded quarterly?
(b) compounded monthly?
(c) compounded continuously?

****

a) $[(1.04)^4 - 1](100) = \underline{16.986\%}$

b) $[(1.01333)^{12} - 1](100) = \underline{17.222\%}$

c) $e^{rn} - 1 = e^{.16(1)} - 1 = \underline{17.35\%}$

2-47 **

If you can make 6% interest on your money, how much is $1000 paid to you 12 years in the future worth to you now?

****

$P = F(P/F, i\%, n) = 1{,}000(P/F, 6\%, 12) = 1{,}000(0.497)$

$= \$497.00$

*** **2-48**

Write the functional notation of a single $2000 payment and evaluate.

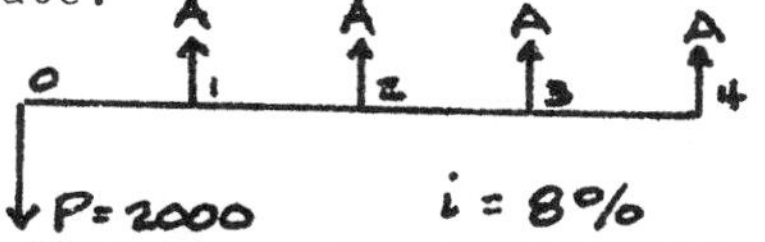

****

$$A = P(A/P, 8\%, 4)$$
$$= \$2{,}000\,(0.3019) = \underline{\$603.80}$$

*** **2-49**

Write the functional notation of the future $500 to determine P, and evaluate.

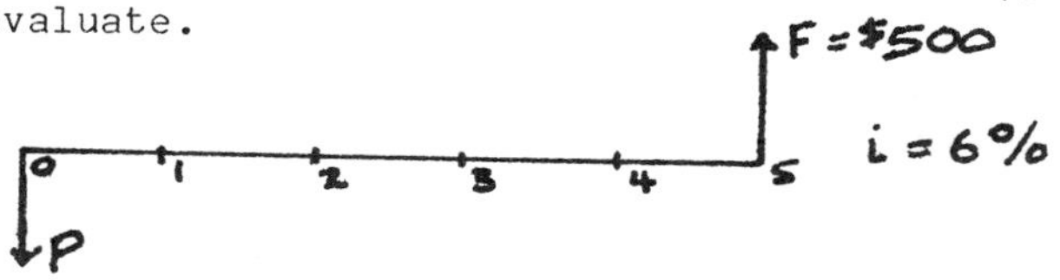

****

$$P = F(P/F, 6\%, 5)$$
$$= \$500\,(0.7473) = \underline{\$373.65}$$

*** **2-50**

Find the Present Equivalent of the following cash flow diagram if i = 18%.

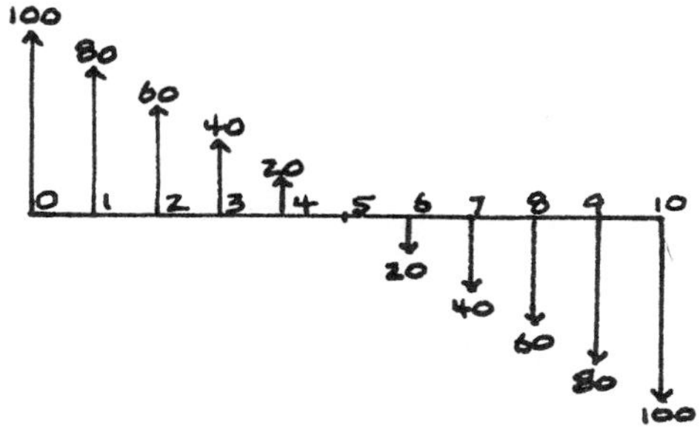

****

$$P = 100 + 80(P/A, 18\%, 10) - 20(P/G, 18\%, 10)$$
$$= 100 + 80(4.494) - 20(14.352) = \underline{\$172.48}$$

2-51 **

Boomtown is experiencing an explosive population growth of 18% per year. At the end of 1984 the population was 16,000. If the growth rate continues unabated, at the end of how many years will the population have passed 75,000?

****

$i = 18\%$

$75{,}000 = 16{,}000(F/P, 18\%, n)$

$(F/P, 18\%, n) = \frac{75{,}000}{16{,}000} = 4.6875$

From table $n = 10$ years

Note that population would not have passed 75,000 after 9 years.

2-52 **

As long as the i and n are the same in all factors, the following relationship always holds exactly.

(F/G,i,n) = (P/G,i,n)(F/P,i,n)

Is the above statement True or False?

****

The statement is True.

2-53 **

To start business, ECON ENGINEERING has just borrowed $500,000 at 6%, compounded quarterly, which will be repaid by quarterly payments of $50,000 each, with the first payment due in one year. How many quarters after the money is borrowed is the loan fully paid off?

****

$i = \frac{6}{4} = 1\tfrac{1}{2}\%$

$P = 500{,}000 = 50{,}000(P/A, 1\tfrac{1}{2}\%, n)(P/F, 1\tfrac{1}{2}\%, 3)$

$(P/A, 1\tfrac{1}{2}\%, n) = \frac{500{,}000}{50{,}000(.9563)} = 10.46$

From tables $n = 12$ payments plus 3 quarters without payments equal 15 quarters before loan is fully paid off.

** **2-54**

Given:

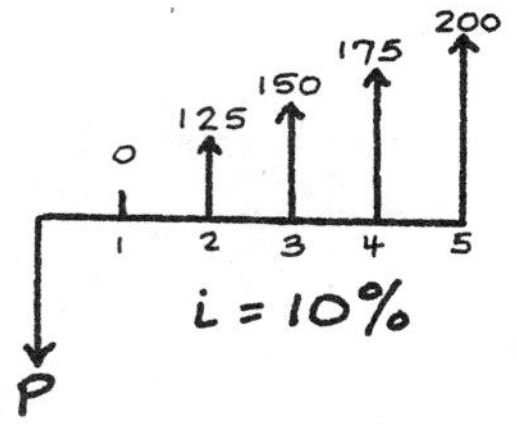

Find: P

****

$P = [125(P/A, 10\%, 4) + 25(P/G, 10\%, 4)](P/F, 10\%, 1)$

$\underline{P = 459.73}$

** **2-55**

Mathematically express:

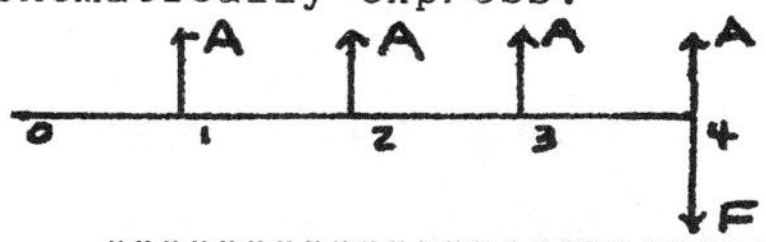

****

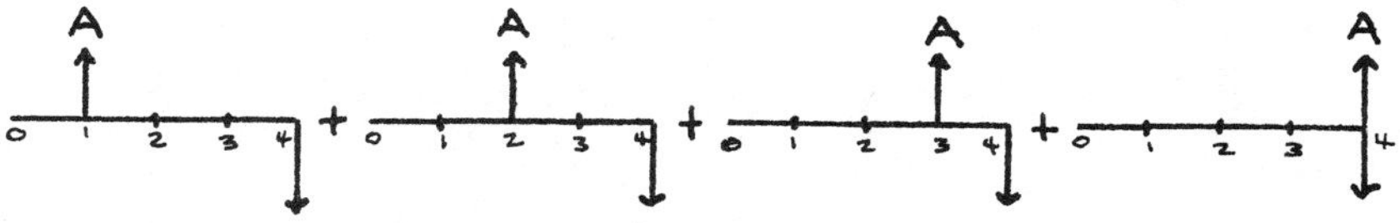

$F = A(1+i)^3 + A(1+i)^2 + A(1+i) + A$

** **2-56**

If the interest rate is 6% compounded quarterly, how long (number of quarters) does it take to earn \$100 interest on an initial deposit of \$300?

****

$i = 6\%/4 = 1\tfrac{1}{2}\%$

$400 = 300(F/P, 1\tfrac{1}{2}\%, n)$

$(F/P, 1\tfrac{1}{2}\%, n) = 400/300 = 1.333$

$\underline{n = 20 \text{ quarters}}$

2-57 **

Draw a diagram that represents (P/F,i%,3).

****

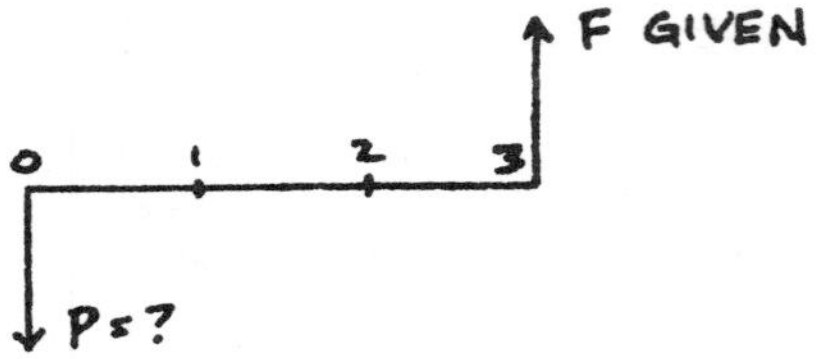

2-58 **

If compounding is weekly and the (one year = 12 months = 48 weeks for this problem) quarterly effective interest rate is 5%,

(a) What is the nominal annual interest rate?
(b) What is the weekly interest rate?
(c) What is the semi-annual effective interest rate?
(d) What is the effective interest rate for a two year period?

****

a) $i_p = (1+r/m)^p - 1 \rightarrow r = m\{\sqrt[p]{1+i_p} - 1\}$

$i_{qr} = i_{12} = .05 = (1 + r/48)^{12} - 1$

$r = 48\{\sqrt[12]{1.05} - 1\} = \underline{19.555\ldots\%}$ per yr. comp wky.

b) $i_{wk} = r/m$

$= 19.555/48 = \underline{.4074\%}$ per week

c) $i_{SA} = i_{24} = i_p = (1+i)^p - 1$

$= (1.004074)^{24} - 1 = \underline{10.25\%}$ per ½ yr.

d) $i_{2yr.} = i_{96} = i_p = (1+i)^p - 1$

$= (1.004074)^{96} - 1 = \underline{47.75\%}$ per 2 yrs.

2-59 **

If the interest rate is 10% compounded continuously, what is the semi-annual effective interest rate?

****

$i_t = e^{rt} - 1$

$i_{1/2} = e^{(.1)(1/2)} - 1 = e^{.05} - 1 = 1.0512711 - 1 = \underline{5.127\%}$ per ½ yr.

3
PRESENT WORTH

** **3-1**

Sarah and her husband decide they will buy $1000 worth of utility stocks beginning one year from now. Since they expect their salaries to increase, they will increase their purchases by $200 per year for the next nine years. What would the present worth of all the stocks be if they yield a uniform dividend rate of 10% throughout the investment period and the price/share remains constant?

****

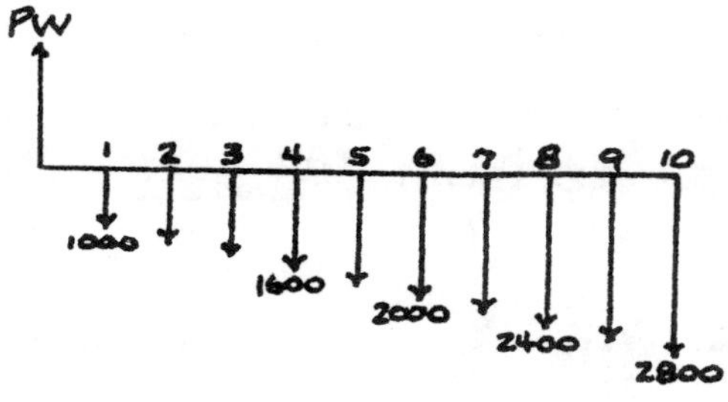

PW of the base amount ($1,000) is:

$$1{,}000\,(P/A, 10\%, 10) = 1{,}000\left[\frac{(1.10)^{10}-1}{(0.10)(1.10)^{10}}\right] = \$\underline{6{,}144.57}$$

PW of the gradient is:

$$200\,(P/G, 10\%, 10) = \frac{200}{0.10}\left[\frac{(1.10)^{10}-1}{0.10} - 10\right]\left[\frac{1}{(1.10)^{10}}\right] = \$4{,}578.27$$

$$\text{Total PW} = 6{,}144.57 + 4{,}578.27 = \$\underline{10{,}722.84}$$

3-2 **

Using an interest rate of 8%, what is the capitalized cost of a tunnel to transport water through the Lubbock mountain range if the first cost is $1,000,000 and the maintenance costs are expected to occur in a 6-year cycle as shown below?

End of Year:	1	2	3	4	5	6
Maintenance:	$35,000	$35,000	$35,000	$45,000	$45,000	$60,000

****

Capitalized Cost = PW of Cost for an infinite time period. As the inital step, compute the Equivalent Annual Maintenance Cost.

$$EUAC = 35{,}000 + [10{,}000(F/A, 8\%, 3) + 15{,}000](A/F, 8\%, 6)$$
$$= 35{,}000 + [10{,}000(3.246) + 15{,}000](0.1363) = 41{,}468.80$$

For $n = \infty$, $P = A/i$

$$\text{Capitalized Cost} = 1{,}000{,}000 + \frac{41{,}468.80}{0.08} = 1{,}518{,}360.$$

3-3 **

The investment in a crane is expected to produce profit from its rental as shown below, over the next six years. Assume the salvage value is zero. What is the present worth of the investment, assuming 12% interest?

Year	Profit	Year	Profit
1	$15,000	4	$7,500
2	12,500	5	5,000
3	10,000	6	2,500

****

$$P = 15{,}000(P/A, 12\%, 6) - 2{,}500(P/G, 12\%, 6)$$
$$= 15{,}000(4.111) - 2{,}500(8.930) = \$\underline{39{,}340}.$$

3-4 **

A tax refund expected one year from now has a present worth of $3000 if **i** = 6%. What is its present worth if **i** = 10%?

****

$$PW = 3{,}000(F/P, 6\%, 1)(P/F, 10\%, 1) = 3{,}000(1.06)(.9091) = \$\underline{2{,}890.94}$$

(let x = amount in year 1 = $3{,}000(F/P, 6\%, 1)$; $PW_x = x(P/F, 10\%, 1)$)

** **3-5**

For the diagram below and an 18% interest rate, compute the value of X so that the two series are equivalent.

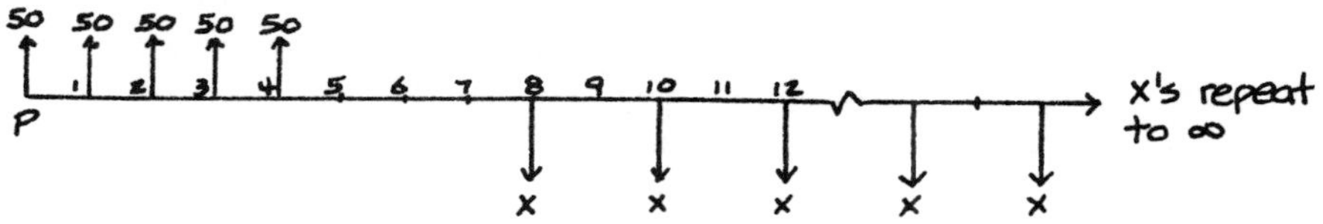

****

Present Equivalent at P $= 50 + 50(P/A, 18\%, 4)$

$= X(A/P, 18\%, 2)(P/A, 18\%, \infty)^{*}(P/F, 18\%, 8)$

$50 + 50(2.690) = X(.6387)\left(\frac{1}{0.18}\right)(.2660)$

$X = \$195.47$

* Note that when $n = \infty$, then $A = Pi$ so $(P/A, i\%, \infty) = \frac{1}{i}$

** **3-6**

An engineer is considering buying a life insurance policy for his family. He currently owes about $77,500 in different loans, and would like his family to have an annual available income of $35,000 indefinitely (that is, the annual interest should amount to $35,000 so that the original capital does not decrease).

(a) He feels he can safely assume that the family will be able to get a 7% interest rate on that capital. How much life insurance should he buy?

(b) If he now assumes the family can get a 10% interest rate, calculate again how much life insurance should be buy.

****

a) If they get 7% interest rate:

$n = \infty$

then $A = Pi$ or $P = A/i = 35{,}000/0.07 = 500{,}000$

Total life insurance $= 77{,}500 + 500{,}000 = 577{,}500$

b) If they can get 10% interest rate:

again $n = \infty$

$P = A/i = 35{,}000/0.10 = 350{,}000$

Total life insurance $= 77{,}500 + 350{,}000 = 427{,}500$

3-7 **

Find the Present Equivalent of the following cash flow diagram if i = 18%.

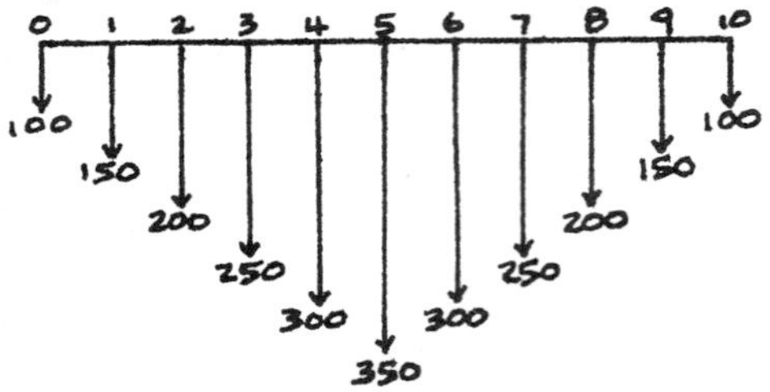

****

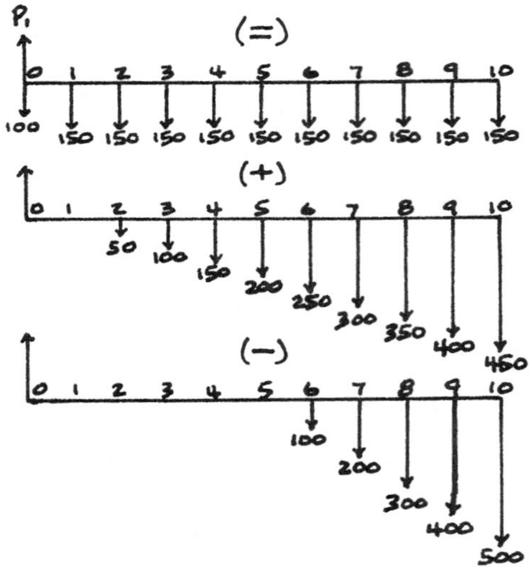

$P_1 = 100 + 150(P/A, 18\%, 10)$
$= 100 + 150(4.494) = 774.10$

$P_2 = 50(P/G, 18\%, 10)$
$= 50(14.352) = 717.60$

$P_3 = 100(P/G, 18\%, 6)(P/F, 18\%, 4)$
$= 100(7.083)(.5158)$
$= 365.34$

$P = P_1 + P_2 - P_3 = \$1{,}126.36$

3-8 **

It takes a full $10,000 to put on a Festival of Laughingly Absurd Works each year. Immediately before this year's FLAW, the sponsoring committee finds that it has $60,000 left in an account paying 15% interest. After this year, how many more FLAWs can be sponsored without raising more money? Think Carefully!

****

$60{,}000 - 10{,}000 = 10{,}000(P/A, 15\%, n)$

$(P/A, 15\%, n) = \frac{50{,}000}{10{,}000} = 5.$

Therefore n = 9 which is the number of FLAWs after this year's. There will be some money left over but not enough to pay for a 10th year.

*** **3-9**

A couple wants to begin saving money for their child's education. They estimate that $10,000 will be needed on the child's 18th birthday, $12,000 on the 19th birthday, $14,000 on the 20th birthday, and $16,000 on the 21st birthday. Assume an 8% interest rate with only annual compounding. The couple is considering two methods of setting aside the needed money.

(a) How much money would have to be deposited into the account on the child's first birthday (note: a child's "first birthday" is celebrated one year after the child is born) to accumulate enough money to cover the estimated college expenses?

(b) What uniform annual amount would the couple have to deposit each year on the child's first through seventeenth birthdays to accumulate enough money to cover the estimated college expenses?

****

a)

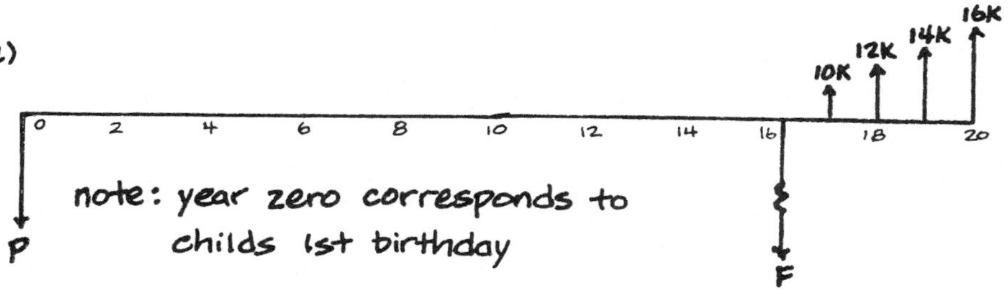

$F = 10{,}000(P/A, 8\%, 4) + 2{,}000(P/G, 8\%, 4)$

$= 10{,}000(3.312) + 2{,}000(4.650) = 42{,}420$

$P = 42{,}420(P/F, 8\%, 16) = 42{,}420(.2919) = \underline{12{,}382.40}$

b)

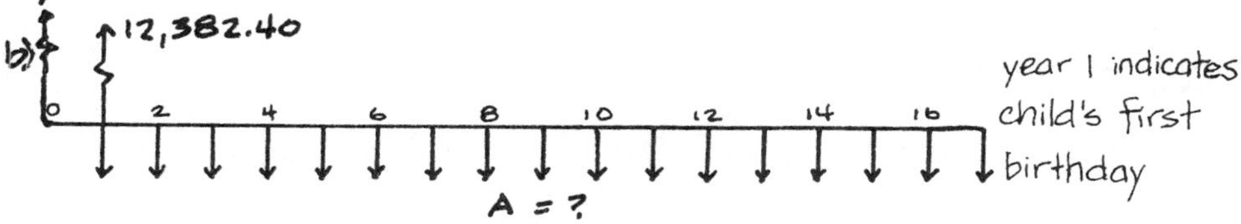

$P' = 12{,}382.40(P/F, 8\%, 1) = 12{,}382.40(.9259) = 11{,}464.86$

$A = 11{,}464.86(A/P, 8\%, 17) = 11{,}464.86(.1096) = \underline{\$1256.55}$

*** **3-10**

Calculate P if P = $3000(P/A,20%,infinity).

****

$P = 3{,}000/.2 = \underline{\$15{,}000}$.

3-11 **

Your company has been presented with an opportunity to invest in a project. The facts on the project are presented below:

Investment required	$60,000,000
Salvage value after 10 years	None
Gross income expected from the project	20,000,000/yr
Operating costs:	
Labor	2,500,000/yr
Materials, licenses, insurance, etc.	1,000,000/yr
Fuel and other costs	1,500,000/yr
Maintenance costs	500,000/yr

The project is expected to operate as shown for ten years. If your management expects to make 25% on its investments before taxes, would you recommend this project?

****

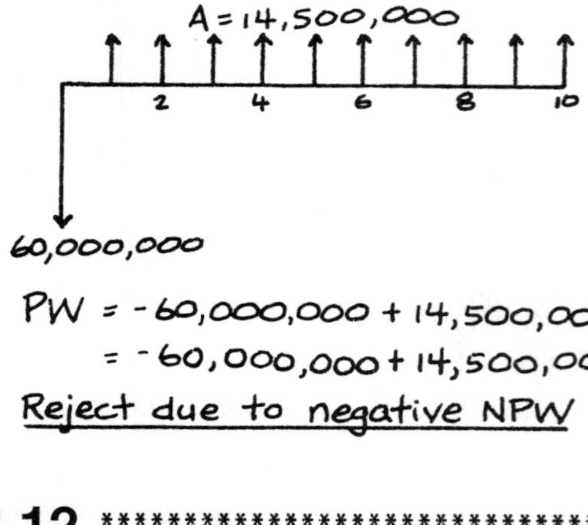

$$PW = -60{,}000{,}000 + 14{,}500{,}000\,(P/A, 25\%, 10)$$
$$= -60{,}000{,}000 + 14{,}500{,}000\,(3.571) = -8{,}220{,}500$$

Reject due to negative NPW

3-12 **

The winner of a sweepstakes prize is given the choice of one million dollars or the guaranteed amount of $81,000 a year for 20 years. If the value of money is taken at a 12% interest rate, which choice is better for the winner?

****

choice 1: $P = \$1{,}000{,}000$

choice 2: $P = 81K\,(P/A, 12\%, 20) = 81K\,(7.469)$

$P = \$604{,}989$

choose alterative 1: take $1,000,000 now

** **3-13**

The annual income from an apartment house is $20,000. The annual expense is estimated to be $2000. If the apartment house could be sold for $100,000 at the end of 10 years, how much could you afford to pay for it now, with 10% considered a suitable interest rate?

****

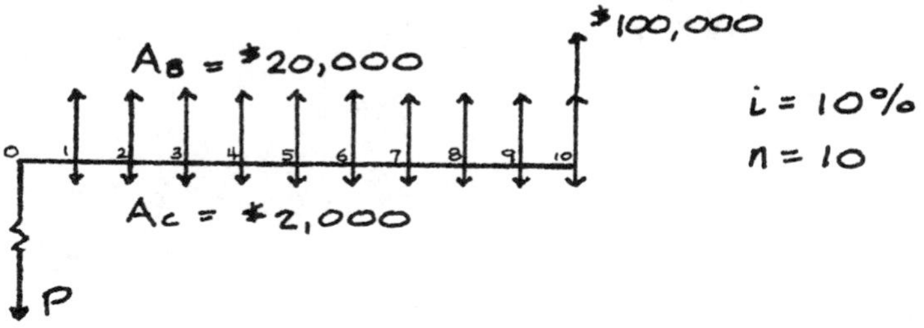

$$P = A_B(P/A, i\%, n) - A_C(P/A, i\%, n) + F_B(P/F, i\%, n)$$
$$= (A_B - A_C)(P/A, i\%, n) + F_B(P/F, i\%, n)$$
$$= (20{,}000 - 2{,}000)(P/A, 10\%, 10) + 100{,}000(P/F, 10\%, 10)$$
$$= 18{,}000(6.145) + 100{,}000(.3855) = \underline{\$149{,}160}$$

** **3-14**

A scholarship is to be established that will pay $200 per quarter at the beginning of Fall, Winter, and Spring quarters. It is estimated that a fund for this purpose will earn 10% interest, compounded quarterly. What lump sum at the beginning of Summer quarter, when deposited, will assure that the scholarship may be continued into perpetuity?

****

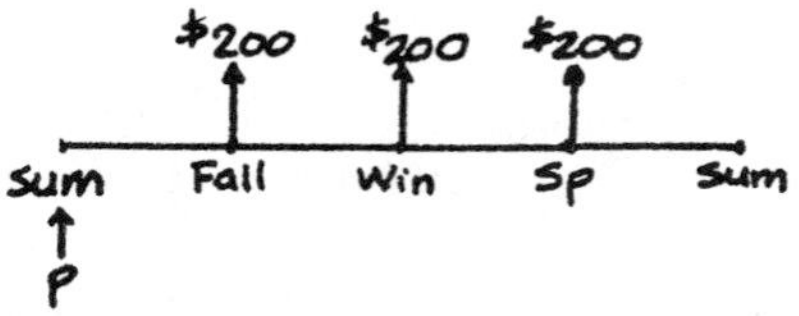

$$P = 200(P/A, 2\tfrac{1}{2}\%, 3)$$
$$= 200(2.856) = 571.20$$
$$A' = 571.20(A/P, 2\tfrac{1}{2}\%, 4)$$
$$= 571.20(.2658) = 151.82$$
For $n = \infty$, $P' = A'/i = 151.82/.025 = \underline{\$6{,}073 \text{ deposit}}$

3-15 **

Using an 8-year analysis and a 10% interest rate, determine which alternative should be selected, based on Net Present Worth (NPW).

Alternative	A	B
First cost	$5,300	$10,700
Uniform Annual Benefit	1,800	2,100
Useful life	4 yrs	8 yrs

****

NPW = PW(benefits) - PW(costs)

alternative A:

$NPW = \$1{,}800\,(P/A, 10\%, 8) - \$5{,}300 - \$5{,}300\,(P/F, 10\%, 4)$

$= 1{,}800\,(5.335) - 5{,}300 - 5{,}300(.6830) = \underline{\$683.10}$

alternative B:

$NPW = \$2{,}100\,(P/A, 10\%, 8) - \$10{,}700 = 2{,}100(5.335) - 10{,}700$

$= \underline{\$503.50}$

select alternative A

3-16 **

Assume you borrowed $50,000 at an interest rate of 1 percent per month, to be repaid in uniform monthly payments for 30 years. In the 163rd payment, how much of it would be interest, and how much of it would be principal?

****

In general, the interest paid on a loan at time t is determined by multiplying the effective interest rate times the outstanding principal just after the preceding payment at time t-1.

To find the interest paid at time t = 163, (call it I_{163}) first find the outstanding principal at time t = 162 (call it P_{162}).

This can be done by computing the future worth at time t = 162 of the amount borrowed, minus the future worth of 162 payments. Alternately, compute the present worth, at time 162, of the 198 payments remaining.

The uniform payments are $50{,}000(A/P, 1\%, 360) = \$514.31$, thus

$P_{162} = 50{,}000(F/P, .01, 162) - 514.31(F/A, 1\%, 162)$

$= 514.31(P/A, 1\%, 198)$

$= \$44{,}259.78$

The interest is $I_{163} = 0.01(44{,}259.78) = \$442.59$

and the principal in the payment is $\$514.31 - 442.59 = \$71.72$

** 3-17**

A municipality is seeking a new tourist attraction, and the town council has voted to allocate $500,000 for the project. A survey shows that an interesting cave can be enlarged and developed for a contract price of $400,000. It would have an infinite life.

The estimated annual expenses of operation are:

Direct Labor	$30,000
Maintenance	15,000
Electricity	5,000

The price per ticket is to be based upon an average of 1000 visitors per month. If money is worth 8 percent, what should be the price of each ticket?

****

If the $100,000 cash, left over after developing the cave, is invested at 8%, it will yield a perpetual annual income of $8000. This $8000 can be used toward the $50,000 a year of expenses. The balance of the expenses can be raised through ticket sales, making the price per ticket

$$\frac{\$42{,}000}{12{,}000 \text{ tickets}} = \$3.50/\text{ticket}$$

Alternate Solution:

PW cost = PW benefit

$$400{,}000 + \frac{(30{,}000 + 15{,}000 + 5{,}000)}{.08} = 500{,}000 + \frac{I}{.08}$$

$$400{,}000 + 625{,}000 = 500{,}000 + \frac{I}{.08}$$

$$I = 525{,}000(.08) = 42{,}000$$

$$\text{Ticket Price} = \frac{42{,}000}{12(1{,}000)} = \$3.50$$

** 3-18**

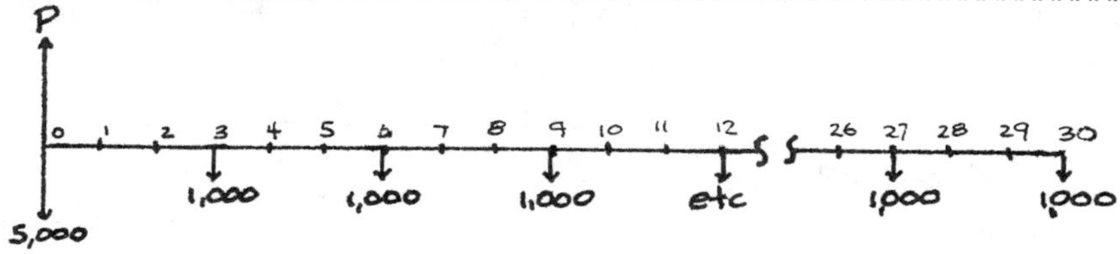

What is the present worth of this initial $5000 investment and $1000 investments every 3 years if i = 12%?

****

$$PW = \$5{,}000 + 1{,}000(A/F, 12\%, 3)(P/A, 12\%, 30)$$

$$= \$5{,}000 + 1{,}000(.2963)(8.055) = \$7{,}387$$

3-19 **

A middle-aged couple has made an agreement with Landscapes Forever Company, a gravesite landscaping and maintenance firm. The agreement states that Landscapes Forever will provide "deluxe landscaping and maintenance" for the couple's selected gravesite forever for an annual fee of $1000. To arrange payment, the couple has set us a variable rate perpetual trust fund with their bank. The bank guarantees that the trust fund will earn a minimum of 5% per year. Assume that the services of Landscapes Forever will not be needed until after the wife has died, and that she lives to the ripe old age of 100.

(a) What is the smallest amount of money that the couple would have to deposit into the trust fund?

(b) Suppose that the couple made this minimum deposit on the wife's 50th birthday, and suppose that the interest rate paid by the trust fund fluctuated as follows:

Wife's Age	Interest Rate
50–54	5%
55–64	10
65–74	15
75–84	20

What is the largest sum of money that could be withdrawn from the trust fund on the wife's 85th birthday, and still have the perpetual payments to Landscapes Forever made?

****

(a) $P = \frac{A}{i} = \frac{1000}{.05} = 20{,}000$

(b)

AGE	i	TRUST FUND BALANCE
50-54	5%	20,000 (F/P, 5%, 5) = 20,000 (1.276) = 25,520
55-64	10%	25,520 (F/P, 10%, 10) = 25,520 (2.594) = 66,198.88
65-74	15%	66,198.88 (F/P, 15%, 10) = 66,198.88 (4.046) = 267,840
75-84	20%	267,840 (F/P, 20%, 10) = 267,840 (6.192) = 1,658,469.43

∴ 1,658,469.43 − 20,000 = $1,632,469.43

3-20 **

Is the following statement True or False?

The capitalized cost is always greater than the present worth of costs for a project of finite life.

****

The statement is True.

** **3-21**

A local car wash charges $3.00 per wash or the option of paying $12.98 for 5 washes, payable in advance with the first wash. If you normally washed your car once a month, would the option be worthwhile if your Minimum Attractive Rate of Return (MARR) is 20% compounded annually?

****

First, convert the effect annual MARR to its equivalent effective monthly rate:

$$(1.2)^{1/12} - 1 = 0.0153$$

Any measure of worth could now be used, but net present value is probably the easiest.

$$NPV = (-12.98 + 3.00) + 3.00(P/A, 1.53\%, 4)$$
$$= 1.57 > 0$$

Therefore, the option is economical.

** **3-22**

A project has a first cost of $10,000, uniform annual benefits of $2400, and a salvage value of $3000 at the end of its 10 year useful life. What is its present worth at an interest rate of 20% per year?

****

$$PW(20\%) = -10{,}000 + (2{,}400)(P/A, 20\%, 10) + 3{,}000\,(P/F, 20\%, 10)$$
$$= 546.45$$

** **3-23**

A person borrows $5000 at an interest rate of 18%, compounded monthly. Monthly payments of $180.76 are agreed upon.

(a) What is the length of the loan?
(Hint: it is an integral number of years.)

(b) What is the total amount that would be required at the end of the sixth month to payoff the entire loan balance?

****

(a) 5,000/180.76 = 27.66, therefore length of loan must be greater than 27.66, try 36.

$A = (5{,}000)(A/P, 1.5\%, 36) = 180.762$, so 36 it is.

(b) $180.762 + (180.762)(P/A, 1.5\%, 30) = \$4{,}521.91$

3-24 **

Given the following three mutually exclusive alternatives.

	Alternative		
	A	B	C
Initial Cost	$50	$30	$40
Annual Benefits	15	10	35
Useful Life (years)	5	5	2

Which alternative is preferable if **i** = 10%?

****

Using simplest method (Net Present Worth):

	A	B	C
Initial Cost	50	30	40
Annual Benefits	15	10	35
Useful Life (years)	5	5	2
Present Worth Benefits	56.865	37.91	60.76
Present Worth Costs	50	30	40
Net Present Worth = PWB - PWC	6.865	7.91	20.76

choose C

3-25 **

A $50,000 30-year loan with a nominal interest rate of 12% compounded monthly is to be repaid in a uniform series of payments of $514.31 per month (for 360 months). The borrower wants to know how many payments, N, he will have to make until he owes only half of the amount he borrowed initially. His Minimum Attractive Rate of Return (MARR) is a nominal 18% compounded monthly.

****

The MARR is irrelevant in this problem. The outstanding principal is always equal to the present worth of the remaining payments when the payments are discounted at the loan's effective interest rate.

Therefore, let N' be the remaining payments.

$\frac{1}{2}(50{,}000) = 514.31(P/A, 10\%, N')$

$(P/A, 10\%, N') = 48.6088$

$N' = 66.9 \approx 67$

So, $N = 360 - N' = 293$ payments.

** **3-26**

Three purchase plans are available for a new car.

Plan A: $5,000 cash immediately
Plan B: $1,500 down and 36 monthly payments of $116.25
Plan C: $1,000 down and 48 monthly payments of $120.50

If a customer expects to keep the car five years and his Minimum Attractive Rate of Return (MARR) is 18% compounded monthly, which payment plan should he choose? Think carefully about the appropriate analysis period.

****

Note that in all cases the car is kept 5 years which is the common analysis period. Therefore PWC is the easiest method.

$i = \frac{18}{12} = 1\frac{1}{2}\%$

$PWC_A = \$5{,}000$

$PWC_B = 1{,}500 + 116.25(P/A, 1\frac{1}{2}\%, 36) = 1{,}500 + 116.25(27.661) = \$4{,}715.59$

$PWC_C = 1{,}000 + 120.50(P/A, 1\frac{1}{2}\%, 48) = 1{,}000 + 120.50(34.043) = \$5{,}102.18$

Therefore Plan B is best

** **3-27**

A project has a first cost of $10,000, net annual benefits of $2000, and a salvage value of $3000 at the end of its 10 year useful life. The project will be replaced identically at the end of 10 years, and again at the end of 20 years. What is the present worth of the entire 30 years of service if the interest rate is 10%?

****

$$\text{PW of 10 years} = -10{,}000 + 2{,}000(P/A, 10\%, 10) + 3{,}000(P/F, 10\%, 10) = \$3{,}445.76$$

$$\text{PW of 30 years} = (3{,}445.76)[1 - (P/F, 10\%, 30)]/[1 - (P/F, 10\%, 10)] = \$5{,}286.45$$

Alternate Solution:

$$\text{PW of 30 years} = [1 + (P/F, 10\%, 10) + (P/F, 10\%, 20)](-10{,}000) + 2{,}000(P/A, 10\%, 30) + 3{,}000[(P/F, 10\%, 10) + (P/F, 10\%, 20) + (P/F, 10\%, 30)] = \$5{,}286.45$$

3-28 **

Consider two investments:

(1) Invest $1000 and receive $110 at the end of each month for the next 10 months.

(2) Invest $1200 and receive $130 at the end of each month for the next 10 months.

If this were your money, and you want to earn at least 12% interest on your money, which investment would you make, if any? What nominal interest rate do you earn on the investment you choose? Solve by Present Worth Analysis.

****

PW Analysis

$i = 12\%/12\text{ mo.} = 1\%$ per month

Alt. 1: $NPW = 110(P/A, 1\%, 10) - 1{,}000 = 110(9.471) - 1{,}000 = +41.81$

Alt. 2: $NPW = 130(P/A, 1\%, 10) - 1{,}200 = 130(9.471) - 1{,}200 = +31.23$

Choose Alt. 1 → Max. NPW

Nominal Interest: $NPW = 0 = -1{,}000 + 110(P/A, i\%, 10)$

$(P/A, i\%, 10) = \frac{1{,}000}{110} = 9.1$

$i \cong 1.75\%$

Nominal interest $= 1.75\% \times 12\text{ mo.} = 21\%$

3-29 **

A farmer has just purchased a tractor for which he had to borrow $20,000. The bank has offered the following choice of payment plans determined using an interest rate of 8%. If the farmer's Minimum Attractive Rate of Return (MARR) is 15%, which plan should he choose?

Plan A: $5,010 per year for 5 years

Plan B: $2,956 per year for 4 years plus $15,000 at end of 5 years.

Plan C: Nothing for 2 years, then $9048 per year for 3 years.

****

$PWC_A = 5{,}010(P/A, 15\%, 5) = 5{,}010(3.352) = \$16{,}794.$

$PWC_B = 2{,}956(P/A, 15\%, 4) + 15{,}000(P/F, 15\%, 5)$

$= 2{,}956(2.855) + 15{,}000(.4972) = \$15{,}897.$

$PWC_C = 9{,}048(P/A, 15\%, 3)(P/F, 15\%, 2) = 9{,}048(2.283)(.7561) = \$15{,}618$

Plan C is lowest cost plan.

** **3-30**

Projects A and B have first costs of $5000 and $9000, respectively. Project A has net annual benefits of $2500 during each year of its 5 year useful life, after which it can be replaced identically.

Project B has net annual benefits of $3300 during each year of its 10 year life. Use present worth analysis, an interest rate of 30% per year, and a 10 year analysis period to determine which project to select.

****

$PW_A = -5{,}000[1+(1.3)^{-5}] + 2{,}500(P/A, 30\%, 10) = \$1{,}382.20$

$PW_B = -9{,}000 + 3{,}300(P/A, 30\%, 10) = \$1{,}202.08$

Select A because of higher present worth of benefits.

** **3-31**

The lining of a chemical tank in a certain manufacturing operation is replaced every 5 years at a cost of $5000. A new type lining is now available which would last 10 years, but costs $13,000. The tank needs a new lining now and you intend to use the tank for 40 years, replacing linings when necessary. With an **i** of 10%, compute the Present Worth of Costs of 40 years of service for the 5-year and 10-year linings.

****

PW 5 yr Lining:

$PW = [5{,}000(A/P, 10\%, 5)](P/A, 10\%, 40)$

$= 5{,}000(.2638)(9.779) = \$12{,}898.50$

PW 10 yr Lining:

$PW = [13{,}000(A/P, 10\%, 10)](P/A, 10\%, 40)$

$= 13{,}000(.1627)(9.779) = \$20{,}683.50$

** **3-32**

The present worth of costs for a $5000 investment with a complex cash flow diagram is $5265. What is the capitalized cost if the project has a useful life of 12 years, and the MARR is 18%?

****

$\text{Capitalized Cost} = 5{,}265(A/P, 18\%, 12)(P/A, 18\%, \infty)$

$= 5{,}265(.2086)(1/.18) = \$6{,}102.$

3-33 ***

A scholarship is to be established to pay $500 each semester to a recipient on September 1 and January 1 of each year. A fund with a yield of 10% will be established on May 1 to pay the proposed scholarship. What deposit at that time will be required to fund the scholarship in perpetuity?

****

Beginning of Summer Semester:

$P' = 500(P/A, \frac{10}{3}\%, 2)$

$= 500(1.904) = 952.13$

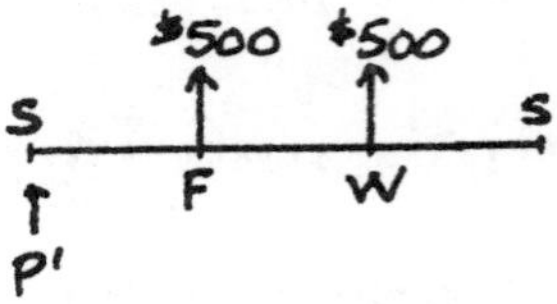

For 3 semesters,

$A' = 952.13(A/P, \frac{10}{3}\%, 3)$

$= 952.13(.356) = 338.77$

$P = \frac{A}{i} = \frac{338.77}{.03333} = \$10{,}163$ to deposit

3-34 ***

A car dealer tells you that if you put $1500 down on a particular car your payments will be $190.93 per month for 4 years at a nominal interest rate of 18%. Assuming monthly compounding, what is the present price you are paying for the car?

****

$A = 190.93$ per period, $i = .18/12 = .015$, $n = 4 \times 12 = 48$

$P = 1{,}500 + A\left[\frac{(1+i)^n - 1}{i(1+i)^n}\right] = 1{,}500 + 190.93\left[\frac{(1+.015)^{48} - 1}{.015(1.015)^{48}}\right]$

$= 1{,}500 + 6{,}499.74 = \$8{,}000$

3-35 ***

What is the price of a 3-year Savings Certificate worth $5000 three years hence, at 12% interest, compounded continuously, with loss of interest if taken out before three years?

****

$P = Fe^{-rn} = \$5{,}000e^{-(0.12)3} = 5{,}000e^{-0.36}$

$= \$5{,}000(0.6977) = \$3{,}488.50$

** **3-36**

If the current interest rate on bonds of a certain type in question is 10 percent nominal, compounded semiannually, what should the market price of a $1000 face value 14 percent bond be? The bond will mature (pay face value) 6-1/2 years from today and the next interest payment to the bondholder will be due in 6 months.

****

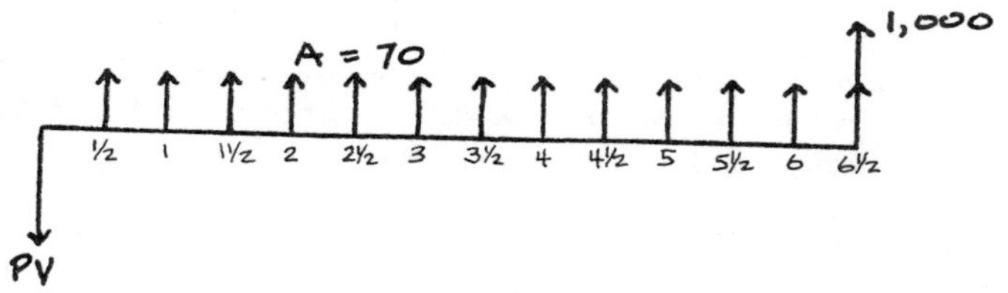

Bi-yearly interest payment = .07(1,000) = \$70

$PV = \$70(P/A, 5\%, 13) + \$1{,}000(P/F, 5\%, 13)$

$= 70(9.394) + 1{,}000(.5303) = \underline{\$1{,}187.90}$

** **3-37**

A manufacturing firm has a before-tax Minimum Attractive Rate of Return (MARR) of 12% on new investments. What uniform annual benefit would Investment B have to generate to make it preferable to Investment A?

	Before Tax Cash Flow	
Year	Investment A	Investment B
0	-$60,000	-$45,000
1-6	+15,000	?

****

NPW of A $= -60 + 15(P/A, 12\%, 6) = 1.665$

NPW of B $\geq 1.665 = -45 + B(P/A, 12\%, 6) \therefore B = 11.351$

$B > \$11{,}351$ per year

** **3-38**

What is the Present Worth of a series that decreases uniformly, by $20 per year, from $400 in Year 11 to $220 in Year 20, if **i** equals 10%?

****

$PW = [400(P/A, 10\%, 10) - 20(P/G, 10\%, 10)](P/F, 10\%, 10)$

$= [400(6.144) - 20(22.8913)](.3855) \doteq \underline{\$770.91}$

3-39 ***

Many years ago BigBank loaned $12,000 to a local homeowner at a nominal interest rate of 4.5%, compounded monthly. The terms of the mortgage called for payments of $60.80 at the end of each month for 30 years. BigBank has just received the 300th payment, thus the loan has five more years to maturity. The outstanding balance is now $3,261.27.

Because BigBank currently charges a nominal 13% compounded monthly on home mortgages, it could earn a better return on its money if the homeowner would pay off the loan now; however, the bank realizes the homeowner has little economic incentive to do that with such a low interest rate on the loan. Therefore, BigBank plans to offer the homeowner a discount.

If the homeowner will pay today an amount of $3,261.27 - D, where D is the dollar amount of the discount, BigBank will consider the loan paid in full. If for BigBank the Minimum Attractive Rate of Return (MARR) is 10% (effective annual rate), what is the maximum discount, D, it should offer the homeowner?

****

The cash flows prior to now are irrelevant. The relevant cash flows are the following:

t	loan continues	paid off early	loan continues minus paid off early
0	0	+(3,261.27-D)	-(3,261.27-D)
1-6	+60.80		+60.80

Any measure of worth could be used. The appropriate discount rate is the effective monthly MARR, $.00797 = (1.1)^{1/12} - 1$

Therefore, using $NPV = 0 = -3261.27 + D + 60.80(P/A, .797\%, 60)$

$D = \$370.60$

3-40 ***

Perpetual care for a small shrine in a cemetary is estimated to be $500/year. The long term interest rate is expected to average 10%. If the capitalized cost is $15,000, what is the first cost of the shrine?

****

$C.C. = 15,000 = P + 500\ (P/A, 10\%, \infty)$

$P = 15,000 - 500\left(\frac{1}{.1}\right) = \underline{\$10,000}.$

** **3-41**

The city council wants the municipal engineer to evaluate three alternatives for supplementing the city water supply. The first alternative is to continue deep well pumping at an annual cost of $10,500. The second alternative is to install an 18" pipeline from a surface reservoir. First cost is $25,000 and annual pumping cost is $7000.

The third alternative is to install a 24" pipeline from the reservoir at a first cost of $34,000 and annual pumping cost of $5000. Life of all alternatives is 20 years. For the second and third alternatives, salvage value is 10% of first cost. With interest at 8%, which alternative should the engineer recommend? Use Present Worth Analysis.

****

Fixed output, minimize cost.

YEAR	DEEPWELL	18" PIPELINE	24" PIPELINE
0		-25,000	-34,000
1-20	-10,500	- 7,000	- 5,000
20		+ 2,500	+ 3,400

Deepwell: PW of Cost = -10,500(P/A, 8%, 20) = -10,500 (9.818)
= -103,089

18" Pipeline: PW of Cost (net of salvage) = -25,000 - 7,000(P/A, 8%, 20) +2,500 (P/F, 8%, 20) = -93,189.75

24" Pipeline: PW of Cost (net of salvage) = -34,000 - 5,000 (P/A, 8%, 20) +3,400 (P/F, 8%, 20) = -82,360.70

Choose 24" Pipeline

** **3-42**

A magazine subscription is $12 annually, or $28 for a 3-year subscription. If the value of money is 12%, which choice is best?

****

$28 \stackrel{?}{=} 12 + 12(P/A, 12\%, 2)$

$28 \stackrel{?}{=} 12 + 12(1.69) = 32.28$

choose 3 yr subscription because 28 < 32.28

3-43 ***

A rich widow decides on her 70th birthday to give most of her wealth to her family and worthly causes, retaining an amount in a trust fund sufficient to provide her with an annual end of year payment of $60,000. If she is earning a steady 10% rate of return on her investment, how much should she retain to provide these payments until she is 95 (the last payment the day before she is 96)? If she dies on her 85th birthday, how much will remain in the trust fund?

****

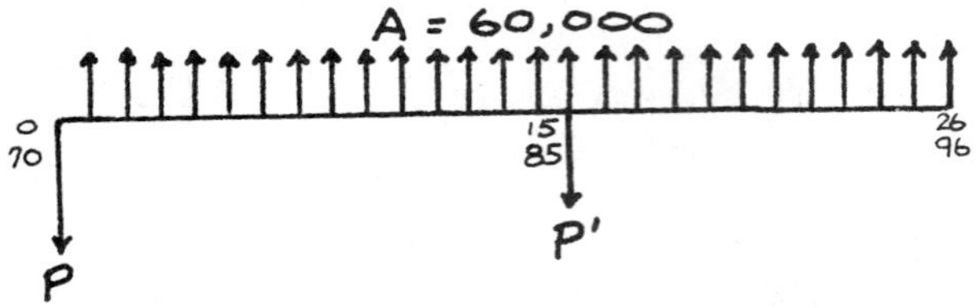

$P = 60K(P/A, 10\%, 26) = 60K(9.161) = \$\underline{549{,}660}.$

$P' = 60K(P/A, 10\%, 11) = 60K(6.495) = \$\underline{389{,}700}.$

3-44 ***

Two alternatives are being considered for recovering aluminum from garbage. The first has a capital cost of $100,000, a first year maintenance cost of $5000, with maintenance increasing by $1500 per year for each year after the first.

The second has a capital cost of $120,000, a first year maintenance cost of $3000, with maintenance increasing by $1000 per year after the first.

Revenues from the sale of aluminum are $20,000 in the first year, increasing $2000 per year for each year after the first. Life of both alternatives is 10 years. There is no salvage value. The before-tax Minimum Attractive Rate of Return is 10%. Using Present Worth Analysis determine which alternative is preferred.

****

Alt. 1: $NPW = -100{,}000 + (20{,}000 - 5{,}000)(P/A, 10\%, 10) + (2{,}000 - 1{,}500)(P/G, 10\%, 10) = \underline{3{,}620.50}$

Alt. 2: $NPW = -120{,}000 + (20{,}000 - 3{,}000)(P/A, 10\%, 10) + (2{,}000 - 1{,}000)(P/G, 10\%, 10) = \underline{7{,}356.00}$

Choose Alt. 2 → Max. NPW

** **3-45**

A brewing company is deciding between two used filling machines as a temporary measure, before a plant expansion is approved and completed. The two machines are:

(a) The Kram Filler. Its initial cost is $85,000, and the estimated annual maintenance is $8000.

(b) The Zanni Filler. The purchase price is $42,000, with annual maintenance costs of $8000.

The Kram filler has a higher efficiency, compared with the Zanni, and it is expected that the savings will amount to $4000 per year if the Kram filler is installed. It is anticipated that the filler machine will not be needed after 5 years, and at that time, the salvage value for the Kram filler would be $25,000, while the Zanni would have little or no value.

Assuming a Minimum Attractive Rate of Return (MARR) of 10%, which filling machine should be purchased?

****

Kram:

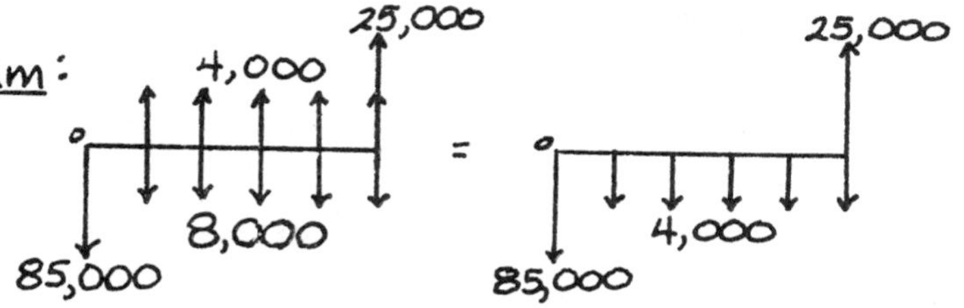

NPW = 25,000(P/F, 10%, 5) - 85,000 - 4,000(P/A, 10%, 5)
= 25,000(.6209) - 85,000 - 4,000(3.791)
= -84,641.5 (or a present worth cost of 84,641.50)

Zanni:

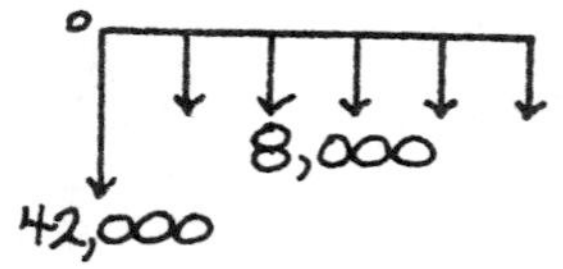

NPW = -42,000 - 8,000(P/A, 10%, 5)
= -42,000 - 8,000(3.791)
= -72,328
(or a PW cost of 72,328)

Therefore choose the Zanni filler.

3-46 **

Two technologies are currently available for the manufacture of an important and expensive food and drug additive. The two can be described as follows:

Laboratory A. Is willing to release the exclusive right to manufacture the additive in this country for \$50,000 payable immediately, and a \$40,000 payment each year for the next 10 years. The production costs are \$1.23 per unit of product.

Laboratory B. This laboratory is also willing to release similar manufacturing rights. They are asking for the following schedule of payments:
On the closing of the contract, \$10,000.
From years 1 to 5, at the end of each year, a payment of \$25,000 each.
From years 6 to 10, also at the end of each year, a payment of \$20,000.
The production costs are \$1.37 per unit of product.

Neither lab is to receive any money after 10 years for this contract. It is anticipated there will be an annual production of 100,000 items for the next 10 years. On the basis of analyses and trials, the products of A and B are practically identical in quality. Assuming a MARR of 12%, which lab should be chosen?

****

Laboratory A: The annual production cost = 1.23 × 100K = \$123K
Cash Flow:

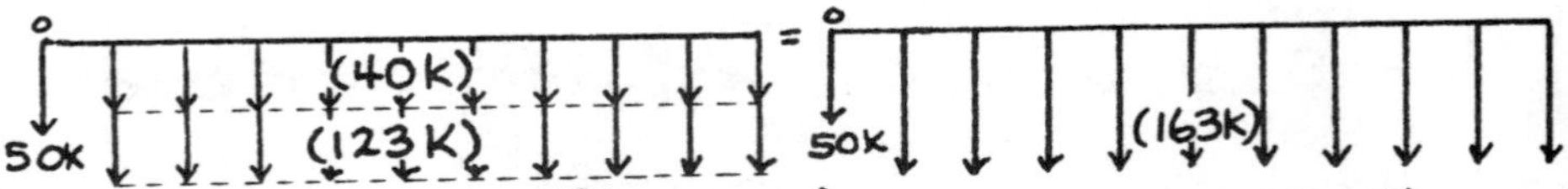

PWC = 50,000 + 163,000(P/A, 12%, 10) = 50,000 + 163,000(5.65)
= 970,950

Laboratory B: The annual production cost = 1.37 × 100K = \$137K

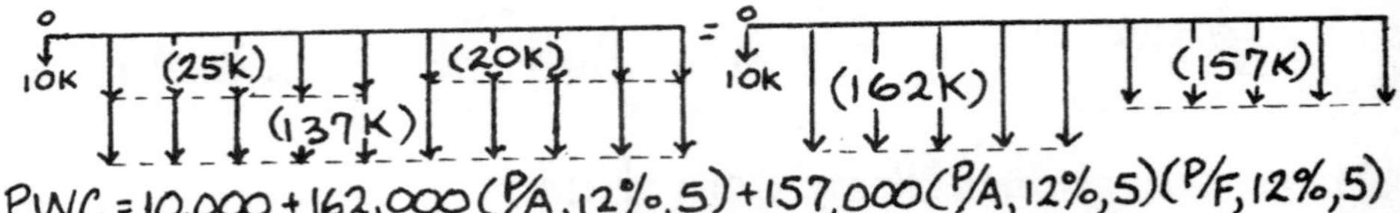

PWC = 10,000 + 162,000(P/A, 12%, 5) + 157,000(P/A, 12%, 5)(P/F, 12%, 5)
= 10,000 + 162,000(3.605) + 157,000(3.605)(.5674) = 915,149.9
Therefore choose Laboratory B.

** 3-47

An engineering analysis by Net Present Worth (NPW) is to be made for the purchase of two devices A and B. If an 8% interest rate is used, recommend the device to be purchased.

	Cost	Uniform Annual Benefit	Salvage	Useful Life
Device A	$600	$100	$250	5 years
Device B	700	100	180	10 "

****

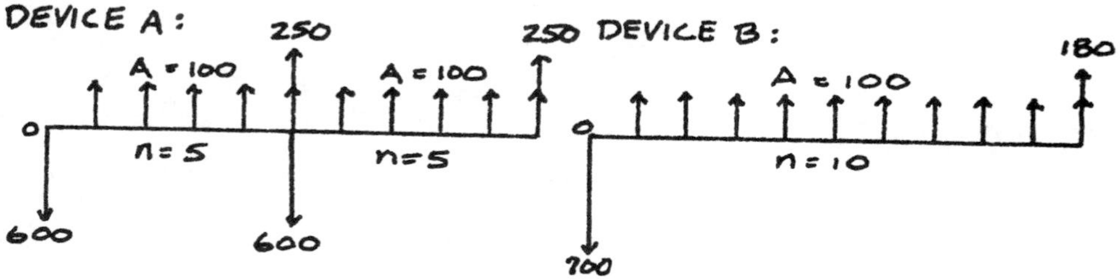

DEVICE A:

$$NPW = 100(P/A, 8\%, 10) + 250(P/F, 8\%, 10) - 600 - [600 - 250](P/F, 8\%, 5)$$

$$= 100(6.710) + 250(.4632) - 600 - 350(.6806)$$

$$= 671 + 115.8 - 600 - 238.21 = \underline{-51.41}$$

DEVICE B:

$$NPW = 100(P/A, 8\%, 10) + 180(P/F, 8\%, 10) - 700$$

$$= 100(6.710) + 180(.4632) - 700 = 671. + 83.38 - 700$$

$$= \underline{+54.38}$$

∴ select device B

** 3-48

A company decides it must provide repair service for the equipment it sells. Based on the following Net Present Worths, which alternative for providing repair service should be selected?

Alternative	Net Present Worth
A	-$15,725
B	-6,657
C	-8,945

****

None of the alternatives look desirable, but since one of the alternatives must be chosen (null not available), choose the one that Maximizes NPW. Thus the best of the three alternatives is B.

3-49 **

A firm is considering the purchase of a new machine to increase the output of an existing production process. Of all the machines considered, the management has narrowed the field to the machines represented by the cash flows shown as follows:

Machine	Initial Investment	Annual Operating Cost
5	\$ 50,000	\$22,500
2	60,000	20,540
4	75,000	17,082
1	80,000	15,425
3	100,000	11,374

If each of these machines provides the same service for 3 years and the minimum attractive rate of return is 12%, which machine should be selected?

****

Minimize the PW of Cost:

Machine	Initial Investment	+ Operating Costs x (P/A, 12%, 3)	= PW of Costs
5	50,000	+ 22,500 (2.402)	= 104,045 ←
2	60,000	+ 20,540 (2.402)	= 109,337
4	75,000	+ 17,082 (2.402)	= 116,031
1	80,000	+ 15,425 (2.402)	= 117,051
3	100,000	+ 11,374 (2.402)	= 127,320

<u>Select Machine 5</u>

3-50 **

J.D. Homeowner has just bought a house with a 30-year, \$30,000 mortgage on which he is paying \$379.33 per month (15% compounded monthly).

(a) If J.D. sells the house after ten years, how much must he give the bank to completely pay off the mortgage at the time of the 120th payment?

(b) How much of the first \$379.33 payment on the loan is interest?

****

a) $P = 379.33 + 379.33(P/A, 1\frac{1}{4}\%, 240)$

$= 379.33 + 379.33(75.942) = \underline{\$29{,}186.41}$

b) $\$30{,}000 \times 0.0125 = \underline{\$375}$

4
ANNUAL CASH FLOW

** **4-1**

While in college Pat received $10,000 in student loans at 5% interest. She will graduate in June and is expected to begin repaying the loans in either 5 or 10 equal annual payments. Compute her yearly payments for both repayment plans.

****

5 YEARS	10 YEARS
$A = P(A/P, i, n)$	$A = P(A/P, i, n)$
$= 10{,}000(A/P, 5\%, 5)$	$= 10{,}000(A/P, 5\%, 10)$
$= 10{,}000(.2310)$	$= 10{,}000(.1295)$
$= \$2{,}310.00$	$= \$1{,}295.00$

** **4-2**

Suppose you wanted to buy a $100,000 house. You have $20,000 cash to use as the down payment. The bank offers to loan you the remainder at 18% nominal interest. The term of the loan is 20 years. Compute your monthly loan payment assuming the payment is the same for all months.

****

$\$100{,}000 - \$20{,}000 = \$80{,}000$ Loan

$A = P(A/P, i\%, n)$

$i = 18\%/12$ periods per yr $= 1.5\%$ per month

$n = 20$ years $\times$ 12 periods per year $= 240$ periods

$A = 80{,}000(A/P, 1.5\%, 240) = 80{,}000(0.0154)$

$A = \$1{,}232.00$ per month

4-3 **

Lester Peabody decides to install a fuel storage system for his farm that will save him an estimated 6.5 cents/gallon on his fuel cost. He uses an estimated 20,000 gallons/year on his farm. Initial cost of the system is $10,000 and the annual maintenance is a uniform gradient amount of $25. After a period of 10 years the estimated salvage is $3000. If money is worth 12%, is it a wise investment?

****

$$EUAC = (P-S)(A/P, i, n) + G(A/G, i, n) + Si$$
$$= (10{,}000 - 3{,}000)(.1770) + 25(3.585) + 3{,}000(.12)$$
$$= 1{,}239 + 89.63 + 360 = \$1{,}688.63$$
$$EUAB = 20{,}000(.065) = \$1{,}300$$

∴ not a wise investment

4-4 **

The returns for a business for five years are as follows: $8,250, $12,600, $9,750, $11,400 and $14,500. If the value of money is 12%, what is the equivalent uniform annual benefit for the five-year period?

****

$$PW = 8{,}250(P/F, 12\%, 1) + 12{,}600(P/F, 12\%, 2) + 9{,}750(P/F, 12\%, 3) + 11{,}400(P/F, 12\%, 4) + 14{,}500(P/F, 12\%, 5)$$
$$= 8{,}250(.8929) + 12{,}600(.7972) + 9{,}750(.7118) + 11{,}400(.6355) + 14{,}500(.5674)$$
$$= \$39{,}823$$
$$EUAB = 39{,}823(A/P, 12\%, 5)$$
$$= 39{,}823(.2774) = \$11{,}047 \text{ Annual Benefit}$$

4-5 **

I have borrowed $1000 from the bank. The interest rate I am to pay is 20%, compounded monthly. I am to repay the loan by making 24 equal monthly payments. What is the amount of my monthly payments?

****

$A = 1{,}000(A/P, \frac{20\%}{12}, 24)$

There are no $1\frac{2}{3}\%$ compounding interest tables readily available. The capital recovery factor must be calculated.

$(A/P, 1.666\%, 24) = \frac{0.01666(1.01666)^{24}}{(1.01666)^{24} - 1} = 0.050892$

$A = 1{,}000(0.050892) = \underline{50.90}$

*** **4-6**

Several companies offer "instant cash" plans to holders of their credit cards. A typical plan permits card holders to "draw" cash up to a preset limit. At the time the cash is drawn, a special charge of 4% of the amount drawn is charged to the card holders account. Then the card holder repays the debt (the original amount drawn plus the special charge) by making a series of equal monthly payments. Each month the company adds a finance charge of 1-1/2% of the previous months unpaid balance to the account balance. If the card holder "draws" $150, a $6 special charge will be made and the card holder will make a series of monthly payments of $9.95.

(a) How many payments will be required?

(b) What "true" (effective) annual interest rate does the card holder pay?

****

(a) $156 = 9.95(P/A, 1\frac{1}{2}\%, n)$

$(P/A, 1\frac{1}{2}\%, n) = 15.678$

From compound interest tables $n = 18$ + a very slight amount

PW of payments $= 9.95(15.673) = 155.95$ for 18 payments

FW of balance $= 0.05(F/P, 1\frac{1}{2}\%, 19) = 0.05(1.327) = 0.07$

So there are 18 payments of $9.95 and a final payment of 7 cents.

(b) $150 = 9.95(P/A, i\%, 18) + 0.07(P/F, i\%, 19)$

solve for i, solution using tables:

try $i = 1\frac{3}{4}\%$ $\quad 150 \stackrel{?}{=} 9.95(15.327) + 0.07(0.7192) = 152.55$

try $i = 2\%$ $\quad 150 \stackrel{?}{=} 9.95(14.992) + 0.07(0.6864) = 149.22$

interpolate: $i = 1\frac{3}{4}\% + \frac{152.55 - 150}{152.55 - 149.22}(\frac{1}{4}\%) = 1.9414\%$

Effective annual interest rate $= (1 + 0.019414)^{12} - 1 = 0.2595$
$= 25.95\%$

Hand calculator solution: $i = 1.940698\%$

Effective annual interest rate $= \underline{25.94\%}$

4-7

****

Data for tractors A and B are listed below. With interest of 12%, which tractor would be selected based on Equivalent Uniform Annual Cost (EUAC)?

	A	B
First cost	$30,000	$36,000
Annual maintenance	1,500	2,000
Salvage value	5,000	8,000
Useful life	6 yrs	6 yrs

****

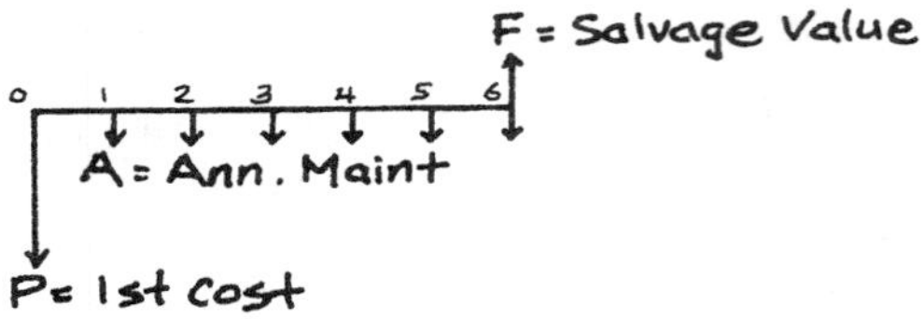

$i = 12\%$

$n = 6$

$\text{EUAC} = P(A/P, i\%, n) - S(A/F, i\%, n) + \text{Other Costs}$

TRACTOR A:

$\text{EUAC} = 30{,}000(A/P, 12\%, 6) - 5{,}000(A/F, 12\%, 6) + 1{,}500$

$= 30{,}000(.2432) - 5{,}000(.1232) + 1{,}500$

$= \$8{,}180$

check: $\text{EUAC} = (P-S)(A/P, i\%, n) + Si + \text{Other Costs}$

$= (30{,}000 - 5{,}000)(.2432) + 5{,}000(.12) + 1{,}500$

$= \$8{,}180$

TRACTOR B:

$\text{EUAC} = 36{,}000(A/P, 12\%, 6) - 8{,}000(A/F, 12\%, 6) + 2{,}000$

$= 36{,}000(.2432) - 8{,}000(.1232) + 2{,}000$

$= \$9{,}769.60$

check: $\text{EUAC} = (P-S)(A/P, 12\%, 6) + S(.12) + \text{Other Costs}$

$= (36{,}000 - 8{,}000)(.2432) + 8{,}000(.12) + 2{,}000$

$= \$9{,}769.60$

Since criteria is to minimize EUAC select tractor A.

4-8

****

If $15,000 is deposited into a savings account that pays 12% interest compounded quarterly, how much can be withdrawn each quarter for five years?

****

$$A = P(A/P, i\%, n)$$
$$= P\left\{\frac{i}{1-(1+i)^{-n}}\right\} = 15{,}000\left\{\frac{.03}{1-(1.03)^{-20}}\right\}$$
$$= 15{,}000(.0672) = \$1{,}008.24 \text{ per quarter}$$

** **4-9**

According to the manufacturers' literature, the costs of running automatic grape peelers, if maintained according to the instruction manuals, are:

Manufacturer:	Slippery	Grater
First Cost	\$500	\$300
Useful Life	10 years	5 years
Maintenance	\$100 at end of years 2, 4, 6 and 8	Year 1 - \$ 0 2 - 50 3 - 75 4 - 100 5 - 125

Which alternative is preferred if MARR = 15%?

****

Slippery:

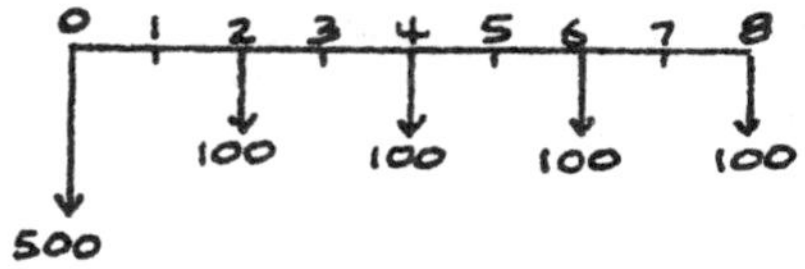

$$EUAC = [500 + 100(A/F, 15\%, 2)(P/A, 15\%, 8)](A/P, 15\%, 10)$$
$$= [500 + 100(.4651)(4.487)](.1993) = \$141.24$$

Grater:

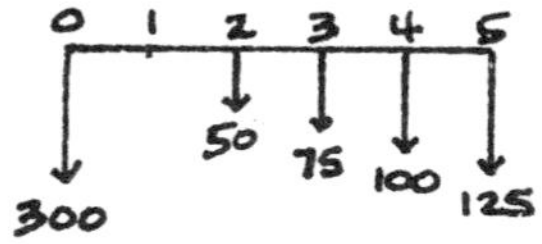

$$EUAC = [300 + 25(P/G, 15\%, 5) + 25(P/A, 15\%, 4)(P/F, 15\%, 1)](A/P, 15\%, 5)$$
$$= [300 + 25(5.775) + 25(2.855)(0.8696)](.2983)$$
$$= \$151.07$$

Therefore, choose Slippery with lower EUAC. (other solutions are possible)

4-10 **

What uniform annual payment for 12 years is equivalent to receiving all of the these:

$ 3,000 at the end of each year for 12 years
20,000 today
4,000 at the end of 6 years
800 at the end of each year forever
10,000 at the end of 15 years

Use an 8% interest rate.

****

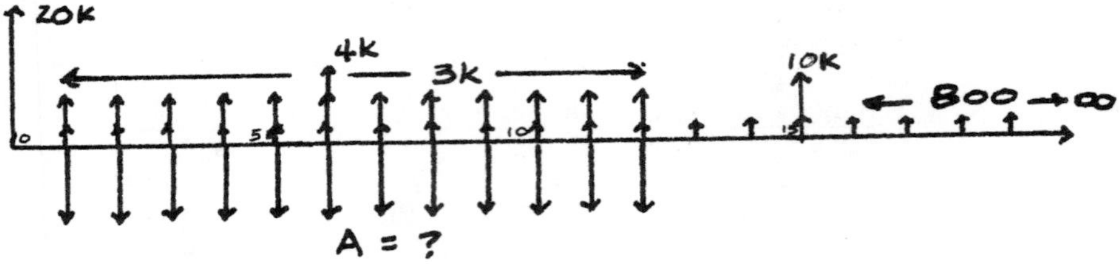

$A_1 = 3{,}000$

$A_2 = 20K(A/P, 8\%, 12) = 20K(0.1327) = 2{,}654$

$A_3 = 4K(P/F, 8\%, 6)(A/P, 8\%, 12) = 4K(.6302)(.1327)$
$= 334.51$

$A_4 = 800/.08(A/P, 8\%, 12) = 10K(0.1327) = 1{,}327$

$A_5 = 10K(P/F, 8\%, 15)(A/P, 8\%, 12) = 10K(.3152)(.1327)$
$= 418.27$

$\sum_{i=1}^{n} A_i = 3{,}000 + 2{,}654 + 334.51 + 1{,}327 + 418.27$
$= \$\underline{7{,}733.78}$

4-11 **

(a) Compute the monthly loan payment on a 30-year loan of $34,000 at 15% interest (nominal) with equal payments.

(b) How much of the first payment is interest, and how much is repayment of principal?

****

a) $A = 34{,}000(A/P, \frac{15}{12}\%, 360) = 34{,}000(0.0126)$
$= 428.40$

b) Interest in first period $= 1.25\%$ of $34{,}000$
$= \underline{425}$.

Principal $= 428.40 - 425 = \underline{3.40}$

** **4-12**

A semiconductor manufacturer has been ordered by the city to stop discharging acidic waste liquids into the city sewer system. Your analysis shows you could select any one of the three systems.

System	Installed Cost	Annual Operating Cost	Salvage value End of 20 yrs
Doxhill	$30,000	$6,000	$ 2,000
Slowsilver	35,000	5,000	5,000
Evergreen	80,000	1,000	40,000

If the system is expected to last and be used 20 years and money is worth 8%, which system should be purchased?

****

Doxhill EUAC $= 6{,}000 + 30{,}000(A/P, 8\%, 20) - 2{,}000(A/F, 8\%, 20)$

$= 6{,}000 + 30{,}000(.1019) - 2{,}000(.0219)$

$= 6{,}000 + 3{,}057 - 44 = \underline{\$9{,}013}$.

Slowsilver EUAC $= 5{,}000 + 35{,}000(.1019) - 5{,}000(.0219)$

$= 5{,}000 + 3{,}565 - 109 = \underline{\$8{,}456}$.

Evergreen EUAC $= 1{,}000 + 80{,}000(.1019) - 40{,}000(.0219)$

$= 1{,}000 + 8{,}152 - 876 = \underline{\$8{,}276}$.

Purchase system with lowest EUAC, Evergreen

** **4-13**

For the following cash flow diagram, which equation properly calculates the Uniform Equivalent (A)?

(a) A = 100(A/P,i,3) + 100(A/F,i,3)
(b) A = 100(A/P,i,15)
(c) A = 100(A/F,i,3) + 100(A/P,i,15)
(d) A = 100(A/F,i,3) + 100(A/F,i,15)
(e) A = 100(A/F,i,3)

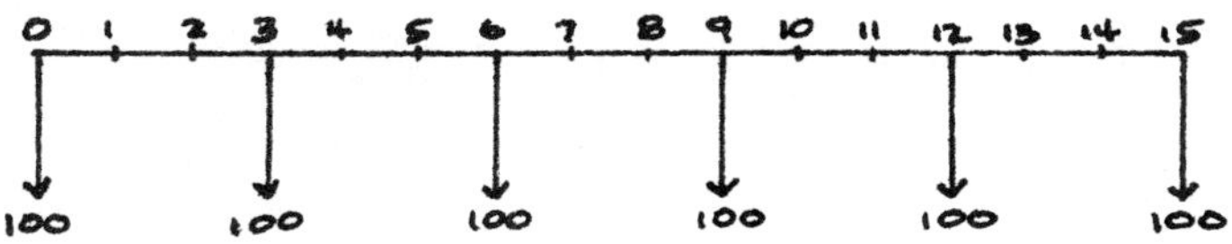

****

The correct equation is (c).

4-14 **

The following alternatives describe possible projects for the use of a vacant lot. In each case the project cost includes the purchase price of the land.

	Parking Lot	Gas Station
Investment Cost	\$50,000	\$100,000
Annual Income	35,000/yr	85,000/yr
Operating Expenses	25,000/yr	70,000 in Year 1, then increasing by 1,000/yr
Salvage	10,000	10,000
Useful Life	5 years	10 years

(a) If the Minimum Attractive Rate of Return (MARR) equals 18%, what should be done with the land?

(b) Is is possible the decision would be different if the MARR were higher than 18%? Why or why not? (No calculations necessary.)

****

a) $NAW_{P.L.} = (35{,}000 - 25{,}000) - 50{,}000(A/P, 18\%, 5) + 10{,}000(A/F, 18\%, 5)$

$= 10{,}000 - 50{,}000(.3198) + 10{,}000(.1398)$

$= \$-4{,}592.$

$NAW_{G.S.} = (85{,}000 - 70{,}000) - 100{,}000(A/P, 18\%, 10) + 10{,}000(A/F, 18\%, 10) - 1{,}000(A/G, 18\%, 10)$

$= 5{,}000 - 100{,}000(.2225) + 10{,}000(.0425) - 1{,}000(3.194)$

$= \$-10{,}019.$

Leave lot vacant

b) No. Higher MARR favors lower cost projects and the lowest cost project (null) has already been choosen.

4-15 **

A project has a first cost of \$75,000, operating and maintenance costs of \$10,000 during each year of its 8 year life, and a \$15,000 salvage value. What is its Equivalent Uniform Annual Cost (EUAC) if the interest rate is 25%?

****

$EUAC(25\%) = (75{,}000)(A/P, 25\%, 8) + 10{,}000 - 15{,}000(A/F, 25\%, 8)$

$= \$31{,}773.91$

** **4-16**

A land surveyor just starting in private practice needs a van to carry crew and equipment. He can lease a used van for $3000 per year, paid at the beginning of each year, in which case maintenance is provided. Alternatively, he can buy a used van for $7000 and pay for maintenance himself. He expects to keep the van three years at which time he could sell it for $1500. What is the most he should pay for uniform annual maintenance to make it worthwhile buying the van instead of leasing it, if his MARR is 20%?

****

Lease:

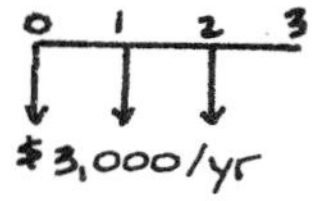

$$EUAC = 3{,}000\,(F/P, 20\%, 1) = 3{,}000\,(1.20)$$
$$= 3{,}600$$

Buy:

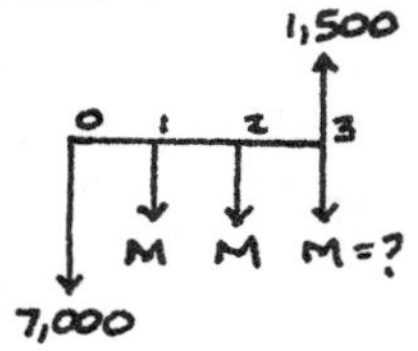

$$EUAC = 7{,}000\,(A/P, 20\%, 3) + M - 1{,}500\,(A/F, 20\%, 3)$$
$$= 7{,}000\,(.4747) + M - 1{,}500\,(.2747)$$
$$= 2910.85 + M$$

$M = 3{,}600 - 2{,}910.85 = \underline{\$689.15}$

** **4-17**

Given the following information about possible investments, what is the best choice at a Minimum Attractive Rate of Return (MARR) of 10%?

	A	B
Investment Cost	$5000	$8000
Annual Benefits	1200	800
Useful Life	5 yrs	15 yrs

****

Net Annual Worth is easier since the useful lives are different

$NAW_A = 1{,}200 - 5{,}000\,(A/P, 10\%, 5) = 1{,}200 - 5{,}000\,(.2638) = \underline{-119.0}$

$NAW_B = 800 - 8{,}000\,(A/P, 10\%, 15) = 800 - 8{,}000\,(.1315) = \underline{-252.0}$

Although A is better than B, the Do-Nothing (Null) alternative is best.

4-18 **

You are considering purchasing the Press-o-matic of Steam-it-out model automatic ironing system to allow you to handle more dry cleaning business. Either machine will cost the same amount, $5000.

The Press-o-matic will generate a positive cash flow of $1300 per year for 5 years and then be of no service or salvage value.

The Steam-it-out will generate a postive cash flow of $800 per year for 10 years and then be of no service or salvage value.

You plan to be in the dry cleaning business for the next 10 years. How would you invest the $5000 you have in your hand if you feel the time value of money is worth the same as your high interest bank account offers, which is

(a) 8%?
(b) 12%?

****

a) Press EUAB = 1,300 − 5,000(.2505) = $47.50
Steam EUAB = 800 − 5,000(.149) = $55.00
choose highest EUAB, steam-it-out.

b) Press EUAB = 1,300 − 5,000(.2774) = −$87
Steam EUAB = 800 − 5,000(.1770) = −$85
choose neither option because both have a cost or negative benefit.

4-19 **

A consumer purchased new furniture by borrowing $1500 using the store's credit plan which charges 18% compounded monthly.

(a) What are the monthly payments if the loan is to be repaid in 3 years?
(b) How much of the first payment is interest?
(c) How much does the consumer still owe just after making the 20th payment?

****

a) $i = 18/12 = 1\tfrac{1}{2}\%$ per month, $n = 3 \times 12 = 36$
$A = 1{,}500\,(A/P, 1\tfrac{1}{2}\%, 36) = 1{,}500(.0362) = \$54.30$

b) Int. = Principal × Monthly intrest rate
Int. = 1,500 × 0.015 = $22.50

c) $P = 54.30\,(P/A, 1\tfrac{1}{2}\%, 16) = 54.30(14.131) = \$767.31$

** **4-20**

A foundation supports an annual seminar on campus by using the earnings of a $50,000 gift. It is felt that 10% interest will be realized for 10 years, but that plans should be made to anticipate an interest rate of 6% after that time. What uniform annual payment may be established from the beginning, to fund the seminar at the same level into infinity?

****

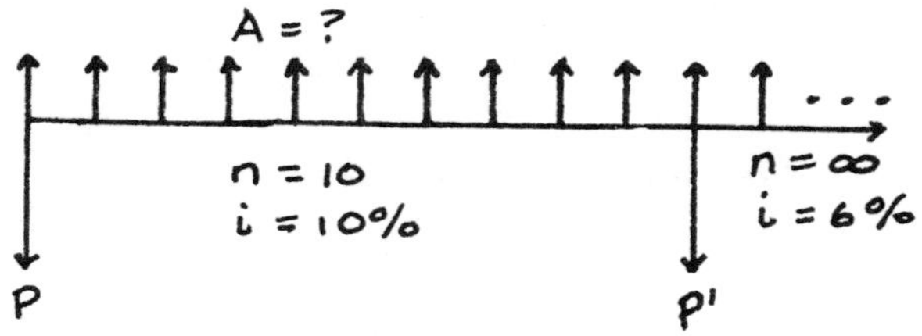

Assume first seminar occurs at time of deposit.

$P' = A/i = A/.06$

$P = A + A(P/A, 10\%, 10) + P'(P/F, 10\%, 10)$

$50{,}000 = A + 6.145A + A/.06 \times .3855$

$13.57A = 50{,}000$

$\underline{A = \$3{,}684.60}$

** **4-21**

A 30-year mortgage of $30,000 at a 12% interest rate had the first payment made on September 1, 1981. What amount of interest was paid for the 12 monthly payments of 1984?

****

Monthly payment $A = 30{,}000\,(A/P, 1\%, 360)$

$= 30{,}000\,(.010286) = 308.58$

Interest periods remaining Jan. 1, 1984 = 331

Jan. 1, 1985 = 319

$P' = 308.58\,(P/A, 1\%, 331)$

$= 308.58\,(96.288) = 29{,}712.51$

$P'' = 308.58\,(P/A, 1\%, 319)$

$= 308.58\,(95.817) = 29{,}567.23$

Interest $= 308.58\,(12) - (29{,}712.51 - 29{,}567.23)$

$= \underline{\$3{,}557.68}$

4-22 **

Data for Machines X and Y are listed below. With an interest of 8%, which machine would be selected based upon Equivalent Uniform Annual Cost (EUAC)?

	X	Y
First cost	$5,000	$10,000
Annual maintenance	500	200
Salvage value	600	1,000
Useful life	5 yrs	15 yrs

****

EUAC = P(A/P, i%, n) − S(A/F, i%, n) + other costs

EUAC = (P−S)(A/P, i%, n) + Si + other costs

Machine X:

EUAC = 5,000 (A/P, 8%, 5) + 500 − 600 (A/F, 8%, 5)

= 5,000 (.2505) + 500 − 600 (.1705)

= 1,252.50 + 500 − 102.3 = $1,650.20

(method 2) EUAC = (5000−600)(A/P, 8%, 5) + 600 (.08) + 500

= 4,400 (.2505) + 48.00 + 500 = $1,650.20

Machine Y:

EUAC = 10,000 (A/P, 8%, 15) + 200 − 1,000 (A/F, 8%, 15)

= 10,000 (.1168) + 200 − 1,000 (.0368)

= 1,168 + 200 − 36.80 = $1,331.20

(method 2) EUAC = (10,000 − 1,000)(.1168) + 1,000 (.08) + 200

= 1,051.20 + 80 + 200 = 1,331.20

Decision criterion, "minimize EUAC, choose Y"

4-23 **

A grateful college graduate makes a donation of $2000 now and will pay $37.50 per month for 10 years to establish a scholarship. If interest in the fund is computed at 9%, what annual scholarship may be established? Assume the first scholarship will be paid at the end of the first year.

****

P = 2,000 + 37.50 (P/A, 3/4%, 120)

= 2,000 + 37.50 (78.942) = 4,960.33

A = 4,960.33 (.09) = $446.43 scholarship

** 4-24**

Consider Projects A and B. Which project would you approve, if the income to both were the same. The expected period of service is 15 years, and the interest rate is 10%.

	Project A	Project B
Initial cost	$50,000	$75,000
Annual operating costs	15,000	10,000
Annual repair costs	5,000	3,000
Salvage value after 15 years	5,000	10,000

****

Project A:

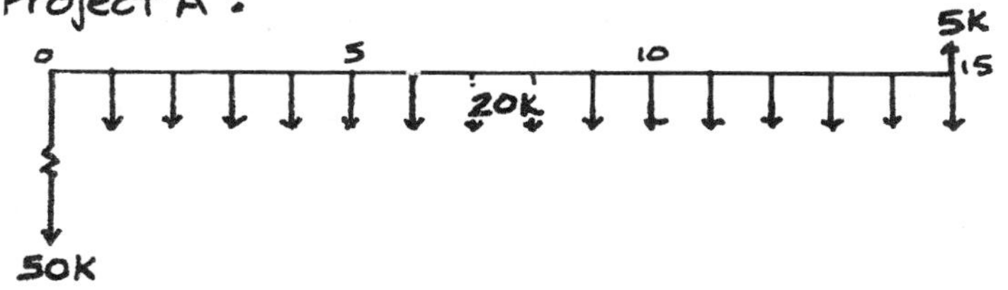

$$EUAC_A = 50K(A/P, 10\%, 15) + 20K - 5K(A/F, 10\%, 15)$$
$$= 50K(.1315) + 20K - 5K(.0315) = 26,417.50$$

Project B:

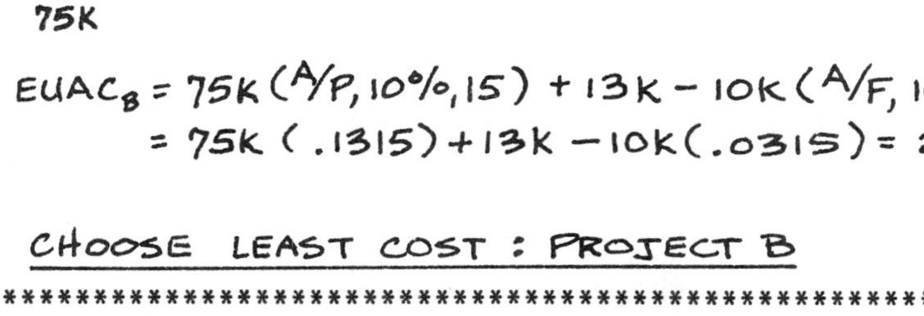

$$EUAC_B = 75K(A/P, 10\%, 15) + 13K - 10K(A/F, 10\%, 15)$$
$$= 75K(.1315) + 13K - 10K(.0315) = 22,547.50$$

CHOOSE LEAST COST : PROJECT B

** 4-25**

A rich folk singer has donated $300,000 to endow a university professorial chair in Bohemian Studies. If the money is invested at 16.5%, how much can be withdrawn each year, ad infinitum, to pay the Professor of B.S.?

****

$$A = 300,000(A/P, 16.5\%, \infty) = 300,000(.165) = \underline{\$49,500}$$

4-26 **

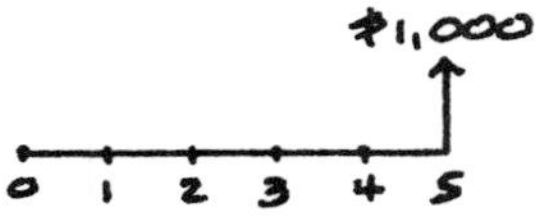

$i = 20\%$

What would an equivalent sum of money be

(a) Now?

(b) Two years from now?

(c) As a five year annuity, starting at the end of the first year?

****

a) $PV = \$1{,}000\,(P/F, 20\%, 5) = \$1{,}000\,(.4019) = \underline{\$401.90}$

b) $\$401.90\,(F/P, 20\%, 2) = \$401.90\,(1.440) = \underline{\$578.74}$

c) $A = \$1{,}000\,(A/F, 20\%, 5) = \$1{,}000\,(.1344) = \underline{\$134.40}$

4-27 **

A project requires an initial investment of $10,000 and returns benefits of $6,000 at the end of every 5th year thereafter. If the Minimum Attractive Rate of Return (MARR) is 10%, should the project be undertaken? Show supporting calculations.

****

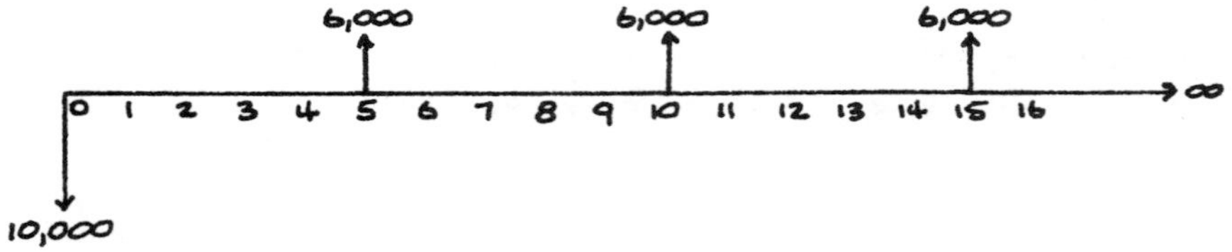

$EUAX = 6{,}000\,(A/F, 10\%, 5) - 10{,}000\,(A/P, 10\%, \infty)$

$= 6{,}000\,(.1638) - 10{,}000\,(0.1000) = \underline{\$-17.20}$

The project should <u>not</u> be undertaken.

4-28 **

On January 1st an engineering student projects a need of $2400 on December 31st. What amount must be deposited in the credit union each month if the interest paid is 12%, compounded monthly?

****

$$A = F\left[\frac{i}{(1+i)^n - 1}\right] = \$2{,}400\left[\frac{.01}{(1.01)^{12} - 1}\right] = \$2{,}400\,(.0788) = \underline{\$189.12}$$

** **4-29**

Given:

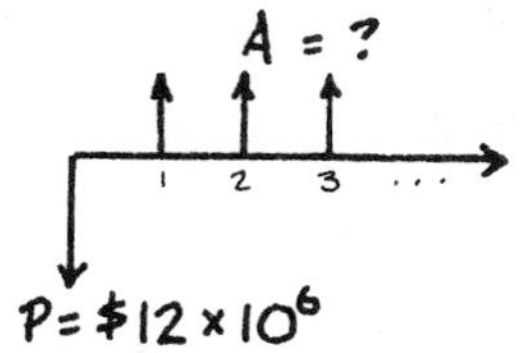

$n = \infty$

$i = 3\%$

Find: A

****

Soln: $A = P \times i$

$A = 12 \times 10^6 \times 0.03$

$\underline{A = 360{,}000}$

** **4-30**

Assuming monthly payments, which would be the better deal on the same $9000 car?

(a) 9% interest on the full amount for 48 months, compounded monthly.

(b) A $1000 rebate (discount) and 15% interest on the remaining amount for 48 months, compounded monthly.

****

a) $A = 9{,}000(A/P, .75\%, 48) = 9{,}000(.0249) = 224.10/\text{mo.}$

b) $A = 8{,}000(A/P, 1.25\%, 48) = 8{,}000(.0278) = 222.40/\text{mo.}$

<u>choose alternative b</u>.

** **4-31**

The initial cost of a pickup truck is $11,500 and will have a salvage value of $4000 after five years. Maintenance is estimated to be a uniform gradient amount of $150 per year (first year maintenance = zero), and the operation cost is estimated to be 30 cents per mile for 300 miles per month. If money is worth 12%, what is the Equivalent Uniform Annual Cost (EUAC) for the truck expressed as a monthly cost?

****

$$\begin{aligned} EUAC &= (11{,}500 - 4{,}000)(A/P, 1\%, 60) + 4{,}000(.01) \\ &\quad + (150(A/G, 12\%, 5))/12 + 300(.30) \\ &= 7{,}500(.0222) + 4{,}000(.01) + (150(1.775))/12 \\ &\quad + 300(.30) = \underline{\$318.69 \text{ per month}} \end{aligned}$$

4-32 **

Twenty five thousand dollars ($25,000) is deposited in a bank trust account that pays 30% interest, compounded semiannually. Equal annual withdrawals are to be made from the account, beginning one year from now and continuing forever. Calculate the maximum amount of the equal annual withdrawal.

****

30% interest compounded semi-annually = $\frac{30\%}{2}$

∴ 15% interest per period

$A = Pi = (25{,}000)(.15) = 3{,}750$

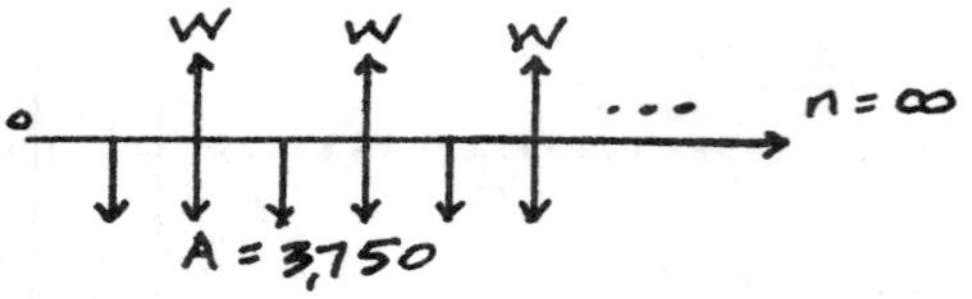

$W = 3{,}750(F/A, 15\%, 2) = 3{,}750(2.150) = 8{,}062.50$

4-33 **

A truck, whose price is $18,600, is being paid for in 36 uniform monthly installments, including interest at 10 percent. After making 13 payments, the owner decides to pay off the remaining balance of the purchase price in one lump sum. How big is this sum?

****

In problems like this the lump sum is the Present Worth of all the future (unpaid) payments. So to solve the problem compute the payment and then compute the PW of the unpaid payments at the stated interest rate.

$A = 18{,}600(A/P, .83\%, 36)$

$= 18{,}600\left[\frac{.00833(1+.00833)^{36}}{(1+.00833)^{36}-1}\right] = \$600.22$

After 13 months : $36 - 13 = 23$

$P = (600.22)(P/A, .83\%, 23) = 600.22(20.85) = \underline{\$12{,}515.45}$

** **4-34**

Assuming a 10% interest rate, determine which alternative should be selected.

	A	B
First Cost	$5,300	$10,700
Uniform Annual Benefit	1,800	2,100
Useful Life	4 years	8 years
Salvage Value	0	200

****

Alternative A:

EUAC = (5,300 − 0)(A/P, 10%, 4) − 1,800
= 5,300 (.3155) − 1,800 = 1,672.15 − 1,800 = $−127.85
= $127.85 benefit

Alternative B:

EUAC = (10,700 − 200)(A/P, 10%, 8) + 200(.1) − 2,100
= 10,500 (.1874) + 200 (.1) − 2,100 = −$112.30
= $112.30 benefit

choose alternative A, higher benefit alternative

** **4-35**

A company must decide whether to buy Machine A or Machine B. After 5 years A will be replaced with another A.

	Machine A	Machine B
First Cost	$10,000	$20,000
Annual Maintenance	1,000	0
End of Useful Life Salvage Value	10,000	10,000
Useful Life	5 years	10 years

With the Minimum Attractive Rate of Return (MARR) = 10%, which machine should be purchased?

****

$EUAX_A$ = −10,000 (A/P, 10%, 5) − 1,000 + 10,000 (A/F, 10%, 5)
= −1,000 − 10,000 i = $−2,000

$EUAX_B$ = −20,000 (A/P, 10%, 10) + 10,000 (A/F, 10%, 10)
= −20,000 (.1627) + 10,000 (.0627) = $−2,627

Therefore Machine A should be purchased.

4-36 **

If the interest rate is 10% and compounding is semiannual, what series of equal annual transactions is equivalent to the following series of semiannual transactions? The first of the equal annual transactions is to occur at the end of the second year and the last at the end of the fourth year.

Time (yr)	0		1		2		3		4		5	5-1/2
Period	0	1	2	3	4	5	6	7	8	9	10	11
Cash Flow	$0	600	500	400	300	200	100	300	500	700	900	1100

****

$P = 600(P/A,5\%,5) - 100(P/G,5\%,5) + [100(P/A,5\%,6) + 200(P/G,5\%,6)](P/F,5\%,5)$

$= 600(4.329) - 100(8.237) + [100(5.076) + 200(11.968)](0.7835)$

$= 4{,}046.8$

Effective $i = (1 + \frac{0.10}{2})^2 = 10.25\%$

Sum at end of Year 1: $F = P(F/P, 10.25\%, 1) = 4{,}046.8(1.1025) = 4{,}461.6$

Equal Annual Payments: $A = 4{,}461.6(A/P, 10.25\%, 3) = 4{,}461.6(0.4039) = 1{,}802.04$

4-37 **

A tractor costs $12,500 and will be used for five years when it is estimated a salvage value of $4000 will be appropriate. Maintenance cost is estimated to be a uniform gradient amount of $100. If a 12% interest rate is used, what is the Equivalent Uniform Annual Cost (EUAC) for the tractor?

****

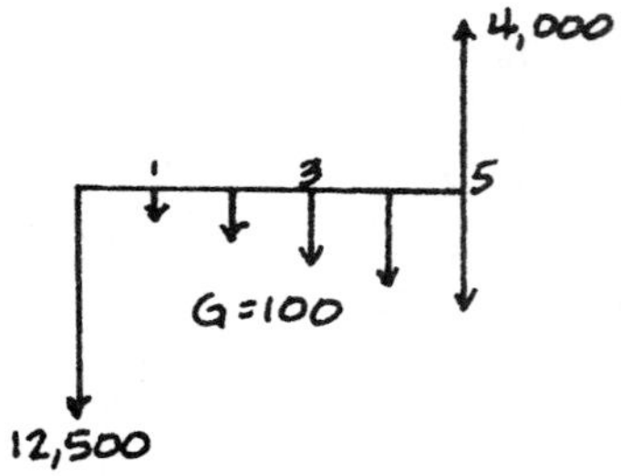

$EUAC = 12{,}500(A/P, 12\%, 5) + 100 + 100(A/G, 12\%, 5)$
$\quad - 4{,}000(A/F, 12\%, 5)$

$= 12{,}500(.2774) + 100 + 100(1.775) - 4{,}000(.1574)$

$= 3{,}115.40$ per year

** **4-38**

I purchased a house for \$50,000 fifteen years ago and I am paying for it over a period of 40 years by making monthly payments. The interest rate is 7%.

(a) How much of my 180th payment is reduction of principal?
(b) How much is interest?
(c) What payment would be necessary if I wanted to pay the loan off at the time I make the 180th payment (do not include the amount of the 180th payment)?

****

(a) Monthly payment (A) = $50{,}000(A/P, \frac{7\%}{12}, 480)$

$= 50{,}000(0.006214) = \$310.72$

Amount owed after 179th payment and before 180th payment

$= 50{,}000(F/P, \frac{7\%}{12}, 179) - 310.72(F/A, \frac{7\%}{12}, 179)$

$= 50{,}000(2.8324) - 310.72(314.1299) = \$44{,}013.56$

Interest portion of 180th payment $= 44{,}013.56 \times \frac{7\%}{12} = 256.75$

Therefore principal portion of 180th payment

$= 310.72 - 256.75 =$ <u>\$53.97</u>

(b) Interest, computed in (a), is <u>\$256.75</u>

(c) Amount owed at 180th payment (not including 180th payment) $= 44{,}013.56 - 53.97 =$ <u>\$43,959.59</u>

** **4-39**

If Zoe won \$250,000 the last week in February, 1984 and invested it by March 1, 1984 in a "sure thing" which pays 8% interest, compounded annually, what uniform annual amount can he withdraw on the first of March for 15 years starting in 1992?

****

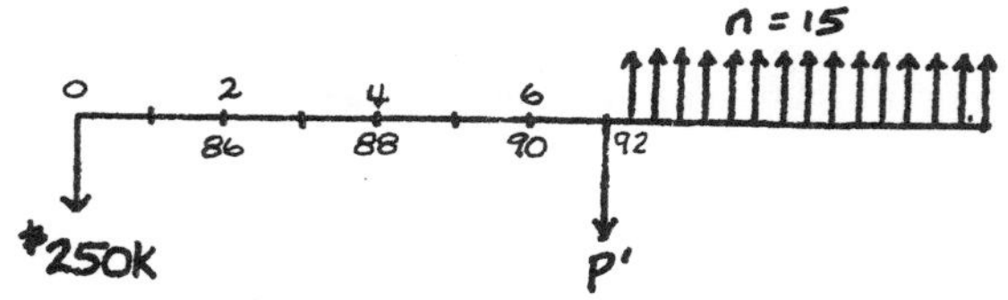

$P' = 250K(F/P, 8\%, 7) = 250K(1.714) = 428{,}500.$

$A = 250K(F/P, 8\%, 7)(A/P, 8\%, 15) = 250K(1.714)(0.1168)$

= <u>\$50,048.80</u>

4-40 **

A machine, with a first cost of $20,000, is expected to save $1,500 in the first year of operation and the savings should increase by $200 every year until (and including) the ninth year, thereafter the savings will decrease by $150 until (and including) the 16th year.

Using Equivalent Uniform Cash Flow (EUCF) as the measure of worth, is this machine economical? Assume a Minimum Attractive Rate of Return of 10%.

****

There are a number of equations that could be written. Here's one:

$$\text{EUCF} = -20{,}000(A/P,10\%,16) + 1{,}500 + 200(P/G,10\%,9)(A/P,10\%,16)$$
$$+ [1{,}450(P/A,10\%,7) - 150(P/G,10\%,7)](P/F,10\%,9)(A/P,10\%,16)$$
$$= -281$$, the machine is not economical

4-41 **

Calculate the equivalent uniform annual cost of the following schedule of payments.

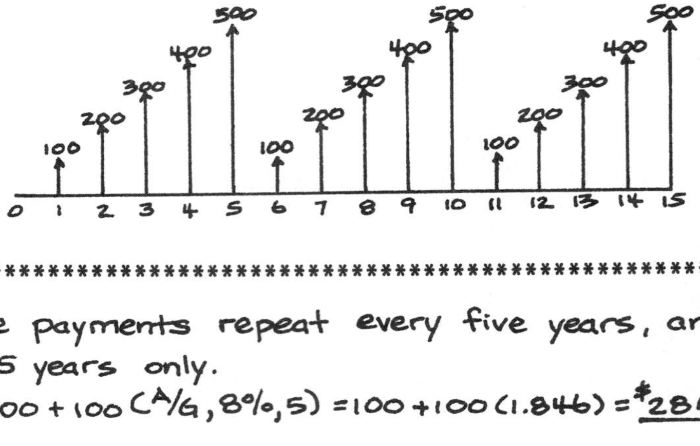

****

Since payments repeat every five years, analyze for 5 years only.

$A = 100 + 100(A/G,8\%,5) = 100 + 100(1.846) = \underline{\$284.60}$

4-42 **

A college is willed $100,000 to establish a permanent scholarship. If funds are invested at 6% and all funds earned are disbursed yearly, what will be the value of the scholarship in the 6th year of operation?

****

$A = Pi = 100{,}000(0.06) = \underline{\$6{,}000}$ for any year

** **4-43**

(a) The construction costs and annual maintenance costs of two alternatives for a canal are given below. Using Equivalent Uniform Annual Cost (EUAC) analysis, which alternative would you recommend? Assume 7% interest and infinite life.

	Alternative A	Alternative B
Construction cost	$25,000,000	$50,000,000
Annual Maintenance costs	3,500,000	2,000,000

(b) What is the capitalized cost of maintenance for the alternative you choose?

****

a)

(A) $EUAC = A + Pi = 3.5 \times 10^6 + 25 \times 10^6 (0.07) = \$5.25 \times 10^6$

(B) $EUAC = 2 \times 10^6 + 50 \times 10^6 (0.07) = \$5.5 \times 10^6$

Fixed Output ∴ minimize cost; choose A

b) $P = \frac{A}{i} = \frac{3{,}500{,}000}{0.07} = \$50 \times 10^6$

** **4-44**

The UNIFORM EQUIVALENT of the cash flow diagram shown is given by which one of the following five answers?

(a) 50(A/G,i,8)
(b) 50(A/G,i,9)
(c) 50(A/G,i,10)
(d) 50(A/G,i,9)(F/A,i,9)(A/F,i,10)
(e) 50(P/G,i,8)(P/F,i,1)(A/P,i,10)

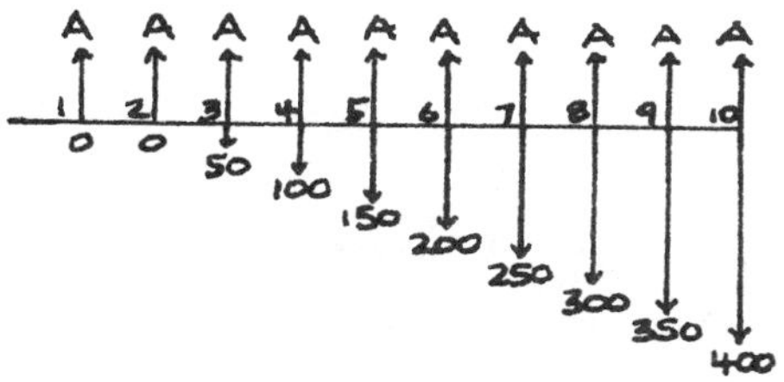

****

Note these two concepts:

1) The G series is 9 periods long

2) The uniform equivalent is 10 periods long

The answer is (d)

4-45 **

Two alternatives are being considered by a food processor for the warehousing and distribution of its canned products in a sales region. These canned products come in standard cartons of 24 cans per carton. The two alternatives are:

Alternative A. To have its own distribution system. The administrative costs are estimated at $43,000 per year, and other general operating expenses are calculated at $0.009 per carton. A warehouse will have to be purchased, which costs $300,000.

Alternative B. To sign an agreement with an independent distribution company, which is asking a payment of $0.10 per carton distributed.

Assume a study period of 10 years, and that the warehouse can be sold at the end of this period for $200,000.

(a) Which alternative should be chosen, if they expect that the number of cartons to be distributed will be 600,000 per year?

(b) Find the minimum number of cartons per year which will make the alternative of having a distribution system (Alt. A) more profitable than to sign an agreement with the distribution company (Alt. B).

****

a) For 600,000 cartons/yr.

Alternative A:

Annual Costs:

Administration:		43,000
Operating Expenses:	.009 × 600,000 =	5,400
Capital Expenses*:		36,270
	Total =	$84,670

$$*\text{EUAC} = (P-S)(A/P, i, n) + Si$$
$$= (300{,}000-200{,}000)(A/P, 10\%, 10) + 200{,}000(0.1)$$
$$= 100{,}000(0.1627) + 20{,}000 = 36{,}270$$

∴ Total annual costs = $84,670.00

Alternative B:

Total annual costs = 0.10 × 600,000 = $60,000

∴ Sign an agreement for distribution (Alt. B)

b) Let M = number of cartons/yr.

- The annual cost for alternative B (agreement) = $EUAC_{agreement} = 0.10M$
- The annual cost for alternative A (own system) = $EUAC_{own} = 43{,}000 + 0.009M + 36{,}270$

We want $EUAC_{own} < EUAC_{agreement}$

$$43{,}000 + .009M + 36{,}270 < 0.10M$$
$$79{,}270 < (0.10 - 0.009)M$$
$$79{,}270/0.091 < M$$
$$871{,}099 < M$$

∴ Own distribution is more profitable for 871,100 or more cartons/year.

** **4-46**

Suppose you purchased a house three years ago by paying 20% down, assuming the first mortgage, and taking a second mortgage (loan) in the amount of $30,000. The term of the second mortgage is 10 years, with a nominal interest rate of 18% compounded monthly. Since that time, interest rates have dropped to 15% on second mortgages. Refinancing the loan will cost you a $100 penalty plus $1000 in new loan fees. Should you refinance the second mortgage if you intend to keep the house for the next seven years?

****

Monthly payment on current loan:

$A = 30{,}000(A/P, \frac{18}{12}\%, 10 \times 12 = 120\text{mo.}) = \$540.$ per month

Remaining principal after 36 monthly payments

$= A(P/A, 1\frac{1}{2}\%, 120 - 36 = 84\text{mo.}) = 540(P/A, 1\frac{1}{2}\%, 84) = 25{,}692.66$

Refinance with penalty and fees:

New loan principal $= 25{,}692.66 + 100 + 1{,}000 = 26{,}792.66$

New monthly payment:

$A = 26{,}792.66(A/P, \frac{15}{12}\%, 84\text{mo.}) = \$517.10$ per month

<u>Refinance the loan</u>

4-47 **

The plant engineer of a major food processing corporation is evaluating alternatives to supply electricity to the plant. He will pay $3 million for electricity purchased from the local utility at the end of this first year and estimates that this cost will increase at $300,000 per year. He desires to know if he should build a 4000 kilowatt power plant. His operating costs (other than fuel) for such a power plant are estimated to be $130,000 per year. He is considering two alternative fuels:

(a) WOOD. Installed cost of the power plant is $1200/kilowatt. Fuel consumption is 30,000 tons per year. Fuel cost for the first year is $20 per ton and is estimated to increase at a rate of $2 per ton for each year after the first. No salvage value.

(b) OIL. Installed cost is $1000/kw. Fuel consumption is 46,000 barrels per year. Fuel cost is $34 per barrel for the first year and is estimated to increase at $1/barrel per year for each year after the first. No salvage value.

If interest is 12%, and the analysis period is 10 years, which alternative should the engineer choose? Solve the problem by Equivalent Uniform Annual Cost analysis (EUAC).

****

	Do Nothing	Wood	Oil
First Cost	0	4,000×1,200 = 4,800,000	4,000×1,000 = 4,000,000
Oper. Cost	0	130,000	130,000
Energy Cost: Annual:	$3,000,000	30,000×20 = 600,000	46,000×34 = 1,564,000
Energy Cost: Gradient:	$ 300,000	30,000×2 = 60,000	46,000×1 = 46,000

Do Nothing:

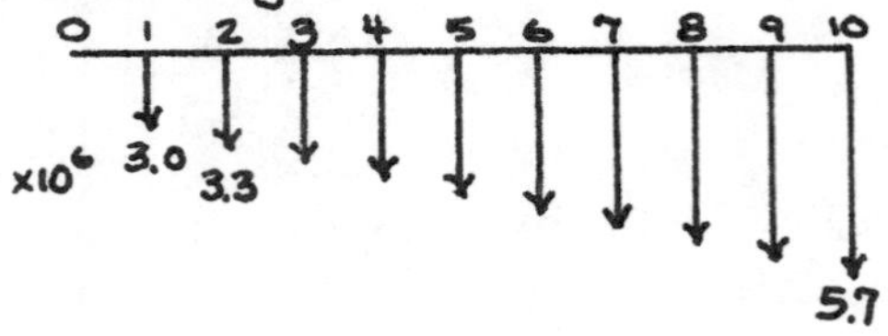

EUAC = 3,000K + 300K (A/G, 12%, 10)
= 3,000K + 300K (3.585)
= $4,075,500 per year

Wood:

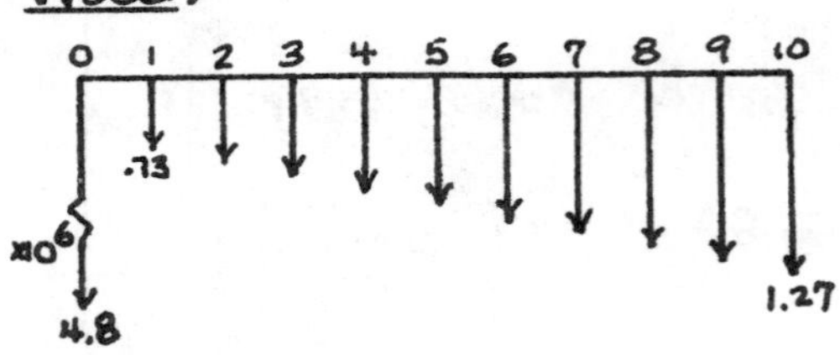

EUAC = 4,800K (A/P, 12%, 10) + 130K
+ 600K + 60K (A/G, 12%, 10)
= $1,794,700 per year

<u>Oil:</u>

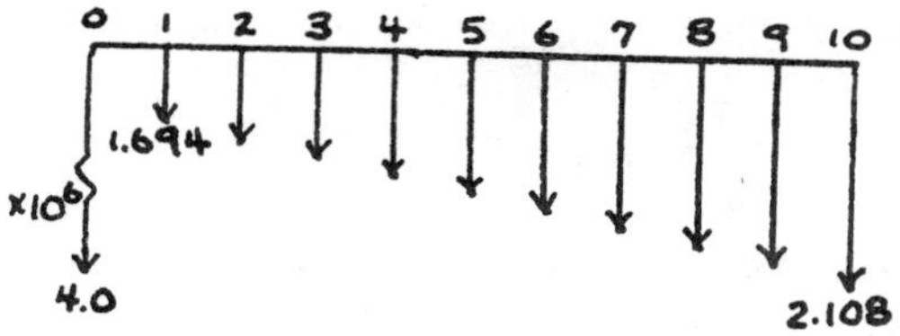

EUAC = 4,000k(A/P,12%,10) + 130k
+1,564k + 46k (A/G,12%,10)
= $2,566,910 per year

<u>Minimize cost ⇒ choose wood</u>

** **4-48**

The manager of F. Roe, Inc. is trying to decide between two alternative designs for an aquacultural facility. Both facilities produce the same number of fish for sale. The first alternative costs $250,000 to build, and has a first-year operating cost of $110,000. Operating costs are estimated to increase by $10,000 per year for each year after the first.

The second alternative costs $450,000 to build, and has a first-year operating cost of $40,000 per year, escalating at $5000 per year for each year after the first. The estimated life of both plants is 10 years and each has a salvage value that is 10% of construction cost.

Assume an 8% interest rate. Using Equivalent Uniform Annual Cost (EUAC) analysis, which alternative should be selected?

****

	Alt. 1	Alt. 2
First Cost	250,000	450,000
Uniform Annual Cost for 10 yrs.	110,000	40,000
Gradient	10,000	5,000
Salvage in year 10	25,000	45,000

EUAC of Alt. 1 = 250,000(A/P, 8%, 10) − 25,000(A/F, 8%, 10) + 110,000
+ 10,000(A/G, 8%, 10) = $184,235 per year

EUAC of Alt. 2 = 450,000(A/P, 8%, 10) − 45,000(A/F, 8%, 10) + 40,000
+ 5,000(A/G, 8%, 10) = $123,300 per year

Fixed Output (same amount of fish for sale) ∴ Minimize EUAC
<u>choose Alt. 2</u>

4-49 **

To get started, Econ Engineering has just borrowed $500,000 which will be paid off in 20 end-of-quarter payments. If interest is 18% compounded monthly, what will be the size of each loan payment?

****

$i = 18\%/12 = 1\frac{1}{2}\%$, $n = 60$ monthly interest periods

$P = 500{,}000 = x(A/F, 1\frac{1}{2}\%, 3)(P/A, 1\frac{1}{2}\%, 60) = x(.3284)(39.380)$

$x = \$38{,}663$

4-50 **

The cost of an automobile is $9000 and after a period of three years it will have an estimated salvage value of $5200. A downpayment of $1000 will be used to purchase the car. It is desired to make the monthly payments, at 12% interest, a value such as to reduce the unpaid balance to exactly the amount of the salvage value after three years. What is the amount of the monthly payment?

****

PW of Salvage $= 5{,}200(P/F, 1\%, 36)$

$= 5{,}200(.6989) = 3{,}634.28$

$P = 9{,}000 - 3{,}634.28 = 5{,}365.72$

$A = 5{,}365.72(A/P, 1\%, 36)$

$= 5{,}365.72(.0332) = \$178.14$ monthly payment

4-51 **

Joyce and Bill purchased a four unit apartment house and as part of the financing obtained a $100,000 loan at 14.5% nominal annual interest, with equal monthly payments for 20 years.

What is their monthly payment?

****

monthly interest $i = \frac{\text{yearly interest}}{12 \text{ months}}$

$i = \frac{.145}{12}$

$i = .01208$ per month

$A = P\left[\frac{i(1+i)^n}{(1+i)^n - 1}\right]$

$= 100{,}000\left[\frac{.01208(1.01208)^{240}}{(1.01208)^{240} - 1}\right] = \$1{,}280.^{00}$ per month

*** **4-52**

The initial cost of a van is $12,800 and will have a salvage value of $5500 after five years. Maintenance is estimated to be a uniform gradient amount of $120 per year, and the operation cost is estimated to be 36 cents/mile for 400 miles/month. If money is worth 12%, what is the Equivalent Uniform Annual Cost (EUAC) for the van, expressed as a monthly cost?

****

$$EUAC = (12{,}800 - 5{,}500)(A/P, 12\%, 5) + (.12)(5{,}500) + 120(A/G, 12\%, 5) + .36(400)(12)$$
$$= 7{,}300(.2774) + .12(5{,}500) + 120(1.775) + 1{,}728$$
$$= (2{,}025 + 660 + 213 + 1{,}728)/12 = \$\underline{385.50/\text{mo.}}$$

*** **4-53**

An engineering student purchased a 2-year-old car that sold new for $8000. The car depreciated 25% per year. The student made a downpayment of $1000 and obtained a 36 month loan at 15% interest, compounded monthly. What were the monthly payments?

****

yr. 1 dep. = (.25)(8,000) = 2,000   8,000 − 2,000 = $6,000

yr. 2 dep. = (.25)(6,000) = 1,500   6,000 − 1,500 = $4,500 sale price

if down payment = 1,000 ; loan = 3,500

$i = 15\%/12 = 1.25\%$; time periods $= n = 12 \times 3 = 36$

$$A = \$3{,}500(A/P, 1\tfrac{1}{4}\%, 36)$$
$$= \$3{,}500(.0347) = \$\underline{121.45}$$

*** **4-54**

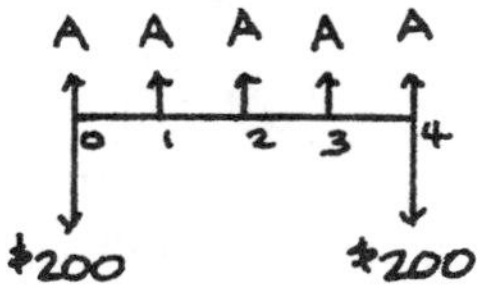

If i = 20%, find A.

****

$$F = \$200 + \$200(F/P, 20\%, 4) = \$200 + \$200(2.074) = 614.80$$
$$A = \$614.80(A/F, 20\%, 5) = \$614.80(.1344) = \$\underline{82.63}$$

4-55 **

An engineer bought a house 4 years ago for $70,000. He paid cash equal to 10% of the purchase price as the downpayment. The rest he financed with two loans. One is a company-subsidized loan of 12% for $20,000, with equal monthly payments for 20 years. The other loan (for the remainder of the money needed) was provided by a local bank, with an interest rate of 15%, also payable over 20 years, with uniform monthly payments. (This local bank charged $1263.00 in various fees, which the engineer paid in cash four years ago.)

The engineer has today (at the end of the fourth year) an option of refinancing both loans with a new loan which has a 10% interest rate, payable in 16 years with uniform monthly payments. The fees for refinancing will amount to $1450.00, and this amount will also have to be borrowed under the same 10% loan.

Should the engineer refinance his loan?

****

a) Find current monthly payments:

Amount needed = 70,000(1 - 0.1) = 63,000

Company Loan = 20,000

Therefore, local bank loan = 63,000 - 20,000 = 43,000

1. Company Loan

$P = 20{,}000$; $n = 20 \times 12 = 240$

$A = P(A/P, i, n) = 20{,}000(A/P, 12\%/12, 240)$

$A = 20{,}000(0.011) = 220.00$

2. Local bank loan*:

$P = 43{,}000$; $n = 240$

$A = 43{,}000(A/P, 15\%/12, 240) = 43{,}000(0.0132) = 567.6$

Therefore, total monthly payment = 220 + 567.6 = <u>787.60</u>

(* the $1263.00 in fees is a sunk cost, not to be included in the analysis)

b) Find the amount still owed today (four years later).

Number of monthly payments left = 16 × 12 = 192

1. Company loan

$$P = 220(P/A, 1\%, 192) = 220\left[\frac{(1+.01)^{192}-1}{0.01(1+.01)^{192}}\right]$$

$$= 220(85.1988) = 18{,}743.74$$

2. Local bank loan

$P = 567.6(P/A, 1.25\%, 192) = 567.6\left[\frac{(1+.0125)^{192}-1}{.0125(1+.0125)^{192}}\right]$

$= 567.6(72.6338) = 41,226.95$

c) Find the amount needed for the new loan:

$18,743.74 + 41,226.95 + 1,450 = 61,420.69$

d) Find the monthly payments under the new loan:

$A = P(A/P, i, n) = 61,420.69(A/P, 10\%/12, 192)$

$= 61,420.69\left[\frac{.00833(1+.00833)^{192}}{(1+.00833)^{192}-1}\right]$

$= 61,420.69(0.010458) = \underline{642.37}$

Therefore, it is worth paying the new loan fee and refinance, since there is a savings of 787.60 − 642.37 = 145.23 \$/month for the next 16 years.

** **4-56**

Two alternative investments are being considered. What is the minimum uniform annual benefit that will make Investment B preferable over Investment A? Assume interest is 10%. Ignore taxes.

Year	A	B
0	−\$500	−\$700
1–5	+150	?

****

$NPW_A = NPW_B$

$-500 + 150(P/A, 10\%, 5) = -700 + x(P/A, 10\%, 5)$

$\frac{200}{(P/A, 10\%, 5)} + 150 = x \quad \Rightarrow x = \frac{200}{3.791} + 150$

$\underline{x = \$202.76}$

Alternate Solution:

$EUAC_A = EUAC_B$

$-500(A/P, 10\%, 5) + 150 = -700(A/P, 10\%, 5) + x$

$200(A/P, 10\%, 5) + 150 = x \quad \Rightarrow x = 200(0.2638) + 150$

$\underline{x = \$202.76}$

4-57 **

A(n) (fill in your major) student bought a car at a used car lot on 4th street for $2000, including tax and insurance. He was to pay for the car by making 19 equal monthly payments, with the first payment to be made when the car was delivered (a down payment). Interest on the loan was charged at the rate of 12% compounded monthly. After 11 payments (the down payment and 10 end-of-month payments) were made, a second buyer agreed to buy the car from the student and to pay a cash amount to pay off the loan in full at the time the next payment was due. If there is no pay off penalty for the early pay off, what amount will be required to pay off the loan?

****

$$P = A(P/A, i\%, n) + A$$

$$2{,}000 = A\left\{\frac{1-(1+i)^{-n}}{i} + 1\right\} = A\left\{\frac{1-(1.01)^{-18}}{.01} + 1\right\}$$

$$= A\{17.398269\}$$

$$A = 2{,}000/17{,}398269 = \underline{114.95397}$$

$$\text{Payoff} = A(P/A, i\%, n) + A = A\left\{\frac{1-(1+i)^{-n}}{i} + A\right\}$$

$$= 114.95397\left\{\frac{1-(1.01)^{-7}}{.01} + 114.95397\right\}$$

$$= 114.95397\{6.7281945 + 1\} = \underline{888.38662}$$

4-58 **

Consider two investments:

(1) Invest $1000 and receive $110 at the end of each month for the next 10 months.

(2) Invest $1200 and receive $130 at the end of each month for the next 10 months.

If this were your money, and you want to earn at least 12% interest on your money, which investment would you make, if any? Solve the problem by annual cash flow analysis.

****

<u>EUAC Analysis</u>:

<u>Alt. 1</u>: EUAB-EUAC $= 110 - 1{,}000(A/P, 1\%, 10) = 110 - 1{,}000(0.1056) = 4.40$

<u>Alt. 2</u>: EUAB-EUAC $= 130 - 1{,}200(A/P, 1\%, 10) = 130 - 1{,}200(0.1056) = 3.28$

<u>Choose Alt. 1 → Max EUAB - EUAC</u>

5
COMPUTING AN UNKNOWN INTEREST RATE

** **5-1**

A large malting plant is considering the installation of energy-saving systems as part of an expansion plan for the plant. The cost of these systems amount to $222,000 and the installation costs represent an additional $178,000. The projected fuel savings are $32,000 per year for the first 5 years of operation, and $53,000 per year from years 6 through 20, when the plant is expected to operate at full capacity. If the MARR of this company is 15%, is this a profitable investment?

****

The cash flow:

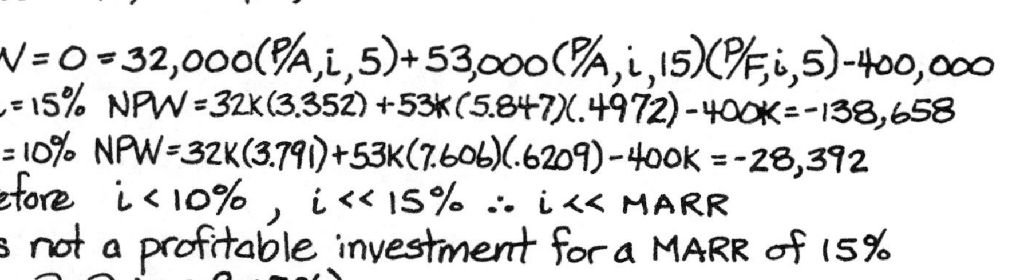

NPW = 0 = 32,000(P/A,i,5) + 53,000(P/A,i,15)(P/F,i,5) − 400,000

for i = 15% NPW = 32K(3.352) + 53K(5.847)(.4972) − 400K = −138,658

for i = 10% NPW = 32K(3.791) + 53K(7.606)(.6209) − 400K = −28,392

Therefore i < 10%, i << 15% ∴ i << MARR

it is not a profitable investment for a MARR of 15%

(The ROR is ≈ 9.07%)

5-2 **

Given: The year-by-year interest rates:

Year	i%
1	14%
2	13
3	12
4	11
5	10

(a) What amount, at the end of five years, is equivalent to $1000 today, given the interest rates shown?

(b) What single interest rate for the same five year period would be equivalent to those in the table?

****

(a) $F = (1.14)(1.13)(1.12)(1.11)(1.10)(1{,}000) = \$1{,}761.16$

(b) $i = (1.7616)^{1/5} - 1 = 0.1199$ or 11.99%. Notice that the result differs slightly from the arithmetic average.

5-3 **

Tony invested $15,000 in a high yield account. At the end of 30 years he closed the account and received $539,250. Compute the effective interest rate he received on the account.

****

$F = P(1+i)^n$

$539{,}250 = 15{,}000(1+i)^{30} \Rightarrow \frac{539{,}250}{15{,}000} = (1+i)^{30}$

$35.95 = (1+i)^{30}$

TRY $i = 12\%$ $\quad 12\% = (1+0.12)^{30} = 29.96$

TRY $i = 13\%$ $\quad 13\% = (1+0.13)^{30} = 39.12$

by interpolation $\quad i = .12 + (.13 - .12)\frac{(35.95 - 29.96)}{(39.12 - 29.96)}$

$i = .126539$ or 12.6539%

5-4 **

Suppose you need to find the value of i% such that (P/A,i%,n) = 5.0000 but you have forgotten the (P/A,i%,n) formula. Fortunately, however, you have two interest tables handy that give (P/A,10%,n) = 2.0000 and (P/A,20%,n) = 22.0000. Use linear interpolation to find i% such that (P/A,i%,n) = 5.0000. (Express your answer as XX.XX%)

****

i%	(P/A, i%, n)
10%	2.000
i%	5.000
20%	22.000

$$i\% = 10\% + 10\%\left(\frac{2-5}{2-22}\right)$$

$$i\% = \underline{11.50\%}$$

** **5-5**

The heat loss through the exterior walls of a processing plant is estimated to cost the owner $3000 next year. A salesman from Superfiber, Inc. claims he can reduce the heat loss by 80% with the installation of $15,000 of Superfiber now.

If the cost of heat loss rises by $200 per year, after next year (gradient), and the owner plans to keep the building ten more years from now, what is his rate of return, neglecting depreciation and taxes?

****

$PW_{cost} = PW_{benefit}$

$15{,}000 = (.8)(3{,}000)(P/A, i\%, 10) + (.8)(200)(P/G, i\%, 10)$

$i = 14\% : 2{,}400\left[\frac{(1.14)^{10}-1}{.14(1.14)^{10}}\right] + \frac{160}{.14}\left[\frac{1.14^{10}-1}{.14} - 10\right]\left[\frac{1}{(1.14)^{10}}\right] = 15{,}397$

$i = 15\% : 2{,}400(5.019) + 160(16.979) = 14{,}762$

by interpolation: $\underline{i = 14.6\%}$

** **5-6**

Does the following project have a positive or negative rate of return? Show how this is known to be true.

Investment cost	$2500
Net benefits	$300 in Year 1, increasing by $200/year
Salvage	$50
Useful life	4 years

****

Year	Benefits
1	300
2	500
3	700
4	900
5	50
Total =	2,450 < Cost

Total Benefits obtained are less than the investment, so the "return" on the investment is negative

5-7

A young engineer has a mortgage loan at a 15% interest rate, which he got some time ago, for a total of $52,000. He has to pay 120 more monthly payments of $620.72 each. As interest rates are going down, he inquires about the conditions under which he could refinance the loan. If the bank charges a new loan fee of 4% of the amount to be financed, and if the bank and the engineer agree on paying this fee by borrowing the additional 4% under the same terms as the new loan, what percentage rate would make the new loan attractive, if the conditions require him to repay it in the same 120 payments?

The amount to be refinanced:

a) PW of 120 monthly payments left

$P = A(P/A, i, n) = 620.72(P/A, \frac{15\%}{12}, 120) = 620.72(61.983) = 38{,}474.09$

b) New loan fee (4%)

$38{,}474.09 \times 0.04 = 1{,}538.96$

$\Rightarrow$ Total amount to refinance $= 38{,}474.09 + 1{,}538.96 = 40{,}013.05$

The new monthly payments are $A_{new} = 40{,}013.05(A/P, i, 120)$

while the current payments are $A_{old} = 620.72$

We want $A_{new} < A_{OLD}$

$40{,}013.05(A/P, i, 120) < 620.72$

then $(A/P, i, 120) < \frac{620.72}{40{,}013.05} = 0.0155$

for $i = 1\%$ $(A/P, 1\%, 120) = 0.0143$

for $i = 1\frac{1}{4}\%$ $(A/P, 1.25\%, 120) = 0.0161$

So, $i \approx 1\% + 0.25\% \left(\frac{0.0143 - 0.0155}{0.0143 - 0.0161}\right) = 1.1667\%$

and this corresponds to a nominal annual percentage rate of $12 \times 0.011667 = 14\%$

Therefore, he/she has to wait until interest rates are smaller than 14%. (at 14% it is practically the same).

5-8

At what interest rate would $1000 at the end of 1985 be equivalent to $2000 at the end of 1992?

$(1+i)^7 = 2$; $i = (2)^{1/7} - 1 = 0.1041$ or 10.41%

** **5-9**

Your company has been presented with an opportunity to invest in a project. The facts on the project are presented below:

Investment required	$60,000,000
Salvage value after 10 years	None
Gross income expected from the project	20,000,000/yr
Operating costs:	
Labor	2,500,000/yr
Materials, licenses, insurance, etc.	1,000,000/yr
Fuel and other costs	1,500,000/yr
Maintenance costs	500,000/yr

The project is expected to operate as shown for ten years. If your management expects to make 25% on its investments before taxes, would you recommend this project?

****

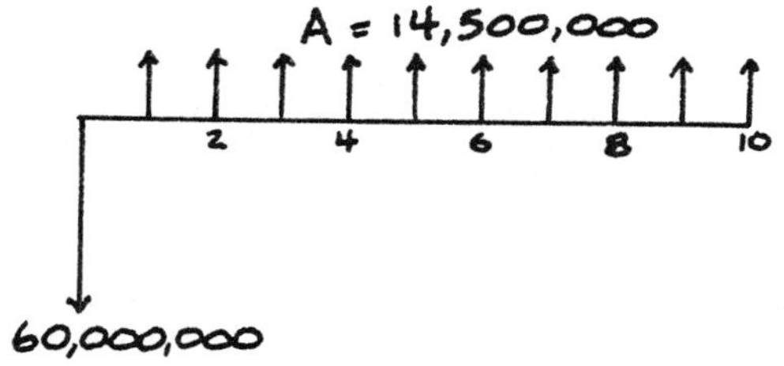

$$-60{,}000{,}000 + 14{,}500{,}000\,(P/A, i, 10)$$
$$(P/A, i, 10) = 60/14.5 = 4.138$$

@ 20%	P/A = 4.192
@ ROR	P/A = 4.138
@ 25%	P/A = 3.571

by interpolation:

$$i = 20\% + 5\left(\frac{4.192 - 4.138}{4.192 - 3.571}\right)$$

$$i = 20.43\%$$

RoR < 25% ∴ Reject Project

** **5-10**

A painting, purchased one month ago for $1000, has just been sold for $1700. What nominal annual rate of return did the owner receive on his investment?

****

$$i = \frac{7{,}000}{1{,}000} = 70\% \qquad ROR = 70 \times 12 = 840\%$$

5-11 **

Consider the following investment in a piece of land.

Purchase price:	\$10,000
Annual maintenance:	\$100/year
Expected sale price after 5 years:	\$20,000

Determine:

(a) A trial value for i

(b) The rate of return (to 1/100 percent)

(c) What is the lowest sale price the investor should accept if she wishes to earn a return of 10% after keeping the land for 10 years?

****

(a)

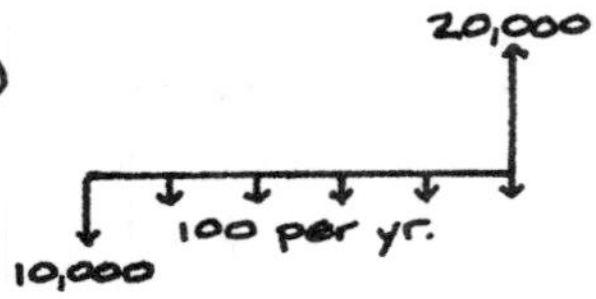

$(F/P, i\%, 5) \approx \frac{20,000}{10,000} = 2$

$\underline{i \approx 15\%}$

(b) $NPW = 20,000(P/F, i\%, 5) - 10,000 - 100(P/A, i\%, 5) = 0$

Try $i = 15\%$: $20,000(.4972) - 10,000 - 100(3.352) = -391.2$

Try $i = 12\%$: $20,000(.5674) - 10,000 - 100(3.605) = +987.5$

$ROR = 12 + 3\left(\frac{987.5}{987.5 + 391.5}\right) = \underline{14.15\%}$

Note: hand calculator answer is 14.10%

(c) $NFW = 0 = \text{Sale Price} - 10,000(F/P, 10\%, 10) - 100(F/A, 10\%, 10)$

$\text{Sale Price} = 10,000(2.594) - 100(15.937)$

$= \underline{\$27,534}.$

5-12 **

Find the rate of return for a \$10,000 investment that will pay \$1000/year for 20 years.

****

$NPW = 1,000(P/A, i\%, 20) - 10,000 = 0$

$(P/A, i\%, 20) = 10$

from tables : $7\% < i < 8\%$ ∴ interpolate

$i = 7\% + 1\%\left(\frac{10 - 10.594}{9.818 - 10.594}\right) = \underline{7.77\%}$

** **5-13**

Calculate the rate of return of the following cash flow with accuracy to the nearest 1/10 percent.

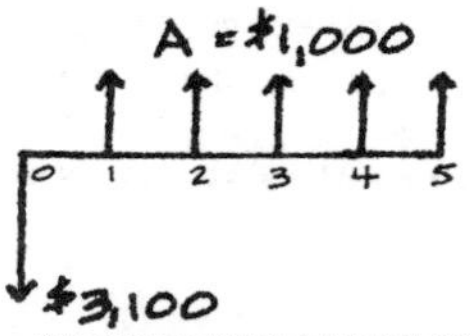

****

$$\$1{,}000\,(P/A, i, 5) = \$3{,}100$$
$$(P/A, i, 5) = 3.1$$

$$(P/A, 20\%, 5) = 2.991$$
$$(P/A, 18\%, 5) = 3.127$$

2% { Δ {	18%	3.127	} .027 } .0136
	x	3.100	
	20%	2.991	

$$\frac{\Delta}{2\%} = \frac{.027}{.0136} \;;\quad \Delta = 2\%\left(\frac{.027}{.0136}\right) = .397$$

$$\therefore\ x = 18\% + .397\% = \underline{18.4\%}$$

** **5-14**

An investment that cost $1000 is sold five years later for $1261. What is the nominal rate of return on the investment (interest rate)?

****

$$F = P\,(F/P, i\%, 5)$$
$$1{,}261 = 1000\,(F/P, i\%, 5)$$
$$(F/P, i\%, 5) = 1261/1{,}000 = 1.2610$$

From tables:

F/P, i%, 5	i%
1.246	4½%
1.261	x
1.276	5%

interpolate to find x

$$x = 4\tfrac{1}{2} + (5 - 4\tfrac{1}{2})\frac{(1.261 - 1.246)}{(1.276 - 1.246)} = 4.75 \text{ or } \underline{4\tfrac{3}{4}\%}$$

5-15 **

Elizabeth made an initial investment of $5000 in a trading account with a stock brokerage house. After a period of 17 months the value of the account had increased to $8400. What is the nominal annual interest rate realized on the initial investment if it is assumed there were no additions or withdrawals from the account?

****

$$F = P(F/P, i, 17)$$

$$\frac{F}{P} = \frac{8{,}400}{5{,}000} = 1.68$$

$$(1+i)^{17} = 1.68$$

$$1+i = (1.68)^{1/17} = 1.031$$

Annual interest rate $= 3.1 \times 12 = 37.2\%$

5-16 **

You borrowed $25 from a friend and after five months repaid $27. What interest rate did you pay (annual basis)?

****

$$F = P(1+i)^n$$

$$\left(\frac{27}{25}\right)^{1/5} = 1+i = 1.0155$$

$i = 1.55\%$ per month or 18.6% per year

5-17 **

You have a choice of $3000 now, or $250 now with $150 a month for two years. What interest rate will make these choices comparable?

****

$$3{,}000 \overset{?}{=} 250 + 150(P/A, i, 24)$$

$P/A = 18.33$

at 2% $P/A = 18.914$

at 2½% $P/A = 17.885$

by interpolation $2\% + .5\left(\frac{18.914 - 18.33}{18.914 - 17.885}\right) = 2.28\%$

or 27.36% per year

6
RATE OF RETURN ANALYSIS

** **6-1**

A large bakery is considering three alternative investments for next year. There were identified by the processing engineering department to be important feasible projects that merit evaluation. They calculated they spent $5600 in the last 6 months gathering the following information:

	Project A Process Automation	Project B Dough-forming Equipment	Project C Improve Packaging
Cost of Equipment	$80,400	$47,800	$75,450
Installation Costs	25,000	20,000	40,000
Expected Annual Saving	22,600	15,100	24,000
Useful life (years)	10	10	10

If MARR is 12%, which alternative should be selected?

****

(The $5600 is a sunk cost, not relevant for comparisons)

1. Calculating the individual ROR's:

For A: $PW_{cost} = PW_{Ben} \Rightarrow$

$(80{,}400 + 25{,}000) = 22{,}600\,(P/A, ROR_A, 10)$

$(P/A, ROR_A, 10) = \frac{105{,}400}{22{,}600} = 4.6637 \Rightarrow ROR_A \approx 17\%\ (>12) \therefore$ attractive

For B: $(47{,}800 + 20{,}000) = 15{,}100\,(P/A, ROR_B, 10)$

$(P/A, ROR_B, 10) = 4.49 \Rightarrow ROR \approx 18\%$ also attractive

For C: $(75{,}450 + 40{,}000) = 24{,}000\,(P/A, ROR_C, 10)$

$(P/A, ROR_C, 10) = 4.81 \Rightarrow ROR \approx 16.2\%$ also attractive

2. Incremental Rates of Return:

$PW_{costs\,B} < PW_{costs\,A} < PW_{costs\,C}$

a) Increment A-B

ΔCosts = (80,400+25,000) − (47,800+20,000) = 37,600

ΔAnn. Ben. = 22,600 − 15,100 = 7,500

$PW_{\Delta costs} = PW_{\Delta Ben.}$ ⇒ 37,600 = 7,500 (P/A, ΔROR_{AB}, 10)

(P/A, ΔROR_{AB}, 10) = 5.013 ⇒ $\Delta ROR_{AB} \approx$ 15% > 12% ∴ keep A

b) Increment C-A

ΔCost = (75,450 + 40,000) − (80,400 + 25,000) = 10,050

ΔAnn. Ben = 24,000 − 22,600 = 1,400

PW (ΔCosts) = PW (ΔBen) ⇒ 10,050 = 1,400 (P/A, ΔROR_{CA}, 10)

(P/A, ΔROR_{CA}, 10) = 7.178 ⇒ $\Delta ROR_{CA} \approx$ 6.5% << 12% ∴ keep A

Select Alternative A.

6-2 ***

A recent graduate wants to join the university's alumni association. There are two payment plans for the association's dues. Plan 1 requires an annual payment of $75. Plan 2 requires a one-time payment of $750. Assume an infinite analysis period and determine the range of values of the Minimum Attractive Rate of Return (MARR) for which Plan 1 is the preferred alternative.

****

YEAR	PLAN 2	PLAN 1	PLAN 2 - PLAN 1
0	-750	-75	-675
1	0	-75	+75

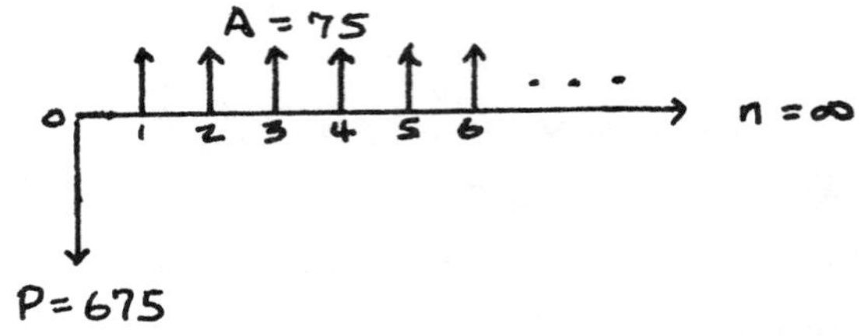

A = Pi

i = A/P = 75/675 = 11.11%

CHOOSE PLAN 1 if MARR > 11.11%

** **6-3**

	Project A	Project B
Initial cost	$50,000	$75,000
Annual operating costs	15,000	10,000
Annual repair costs	5,000	3,000
Salvage value after 15 years	5,000	10,000

If the income to both Projects A and B is the same, what rate of return would an investor make on the additional investment required by choosing Project B rather than Project A?

****

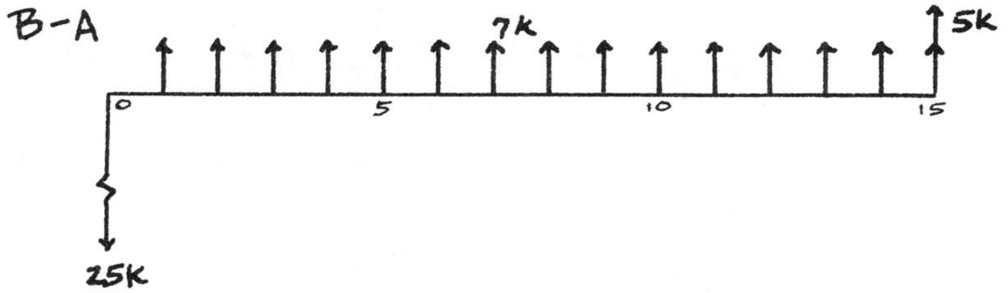

$PW = -25K + 7K(P/A, i, 15) + 5K(P/F, i, 15)$

$P/A = 25/7 = 3.51743$ ∴ from tables $i \approx 30\%$

NPW @ 25% → $-25k + 7k(P/A, 25\%, 15) + 5k(P/F, i, 15)$

$-25k + 7k(3.859) + 5K(.0352) = +2,189.$

NPW @ 30% → $-25k + 7k(P/A, 30\%, 15) + 5k(P/F, 30\%, 15)$

$-25K + 7K(3.268) + 5k(.0195) = -2,026.5$

by interpolation: $i = 25 + 5\left(\frac{2,189}{2,189 + 2,026.5}\right) = 25 + 2.5964$

$i = \underline{27.596\%}$

** **6-4**

Given the two alternatives shown below, which (if either) is preferred at a Minimum Attractive Rate of Return (MARR) of 8%?

	A		B
Investment	$500		$700
Rate of Return	10%		10%
Incremental Rate of Return		10%	

****

Since ROR > MARR, both are acceptable.
Since ΔROR > MARR, higher cost project is preferred.
Therefore <u>B is best</u>.

6-5 **

Given the costs and benefits of two electric motors, with the analysis period of 3 years, what is the rate of return on the difference of these alternatives? Select the nearest whole number of i.

Years	A	B
0	-$2000	-$2800
1	+800	+1100
2	+800	+1100
3	+800	+1100

****

YR	A	B	DIFFERENCE B-A
0	-2,000	-2800	-800
1	+800	+1,100	+300
2	+800	+1,100	+300
3	+800	+1,100	+300

PW(costs) = PW(benefits)

$800 = 300(P/A, i\%, 3)$

$(P/A, i\%, 3) = 800/300 = 2.67$, from tables $i = 6\%$

6-6 **

The following mutually exclusive alternatives have been identified for a possible investment. Make a choice table showing the range of Minimum Attractive Rate of Return (MARR) in which each alternative is preferable. The null (do nothing) alternative is available. Report Rates of Return to the nearest 1/10%. (HINT: Use Net Present Worth; also use i = 10% for the first trial for the incremental rate of return.)

	A	B
First Cost	$4,000	$2,000
Annual Benefit	0	477
Salvage Value	12,424	0
Useful Life	10 years	10 years

****

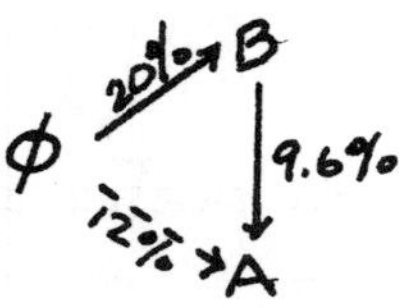

$ROR_A: (P/F, i\%, 10) = \frac{4000}{12,424} = 0.3220$

$i = 12.0\%$

$ROR_B: (P/A, i\%, 10) = \frac{2,000}{477} = 4.193$

$i = 20.0\%$

ΔROR_{A-B}

$NPW_H - NPW_L = [12{,}424(P/F, i\%, 10) - 4{,}000]_A - [477(P/A, i\%, 10) - 2{,}000]_B = 0$

$= 12{,}424\,(P/F, i\%, 10) - 477\,(P/A, i\%, 10) - 2{,}000$

Try $i = 10\%$: $12{,}424(.3855) - 477(6.145) - 2000 = -141.71$

Try $i = 9\%$: $12{,}424\,(.4224) - 477(6.418) - 2{,}000 = 186.51$

$\Delta ROR_{A-B} = 9 + 1\left(\frac{186.51}{141.71 + 186.51}\right) = 9.57\% = \underline{9.6\%}$

choose ϕ if $20\% < MARR$

B if $9.6\% < MARR < 20\%$

A if $MARR < 9.6\%$

*** **6-7**

Given the costs and benefits of two water pumps, what is the rate of return on the difference of these alternatives? Compute to the nearest integer value.

Years	A	B
0	-$3000	-$3800
1	+800	+1200
2	+800	+1200
3	+800	+1200
4	+800	+1200
5	+800	+1200

****

YEAR	A	B	DIFFERENCE B-A
0	-3,000	-3,800	-800
1	+800	+1,200	+400
2	+800	+1,200	+400
3	+800	+1,200	+400
4	+800	+1,200	+400
5	+800	+1,200	+400

PW (costs) = PW (benefits)

$800 = 400\,(P/A, i\%, 5)$

$(P/A, i\%, 5) = 800/400 = 2 \quad \therefore \ \underline{i \approx 40\%}$

6-8 **

A local recreational center offers a one year membership for \$180 and a two year membership for \$270 (a 25% discount compared with two years for \$360). Memberships are paid in advance.

(a) What is the incremental rate of return?
(b) Which membership should be chosen if the center will be used for at least 2 years? Why?

****

a) The cash flow diagram of the difference is:

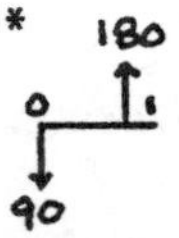

$\Delta ROR = \frac{90}{90} = 100\%$

b) Choose 2-year plan since it seems unlikely anyone has a MARR > 100%

6-9 **

The manager of a local restaurant is trying to decide whether to buy a charcoal broiling unit or an electric grill for cooking hamburgers. A market study shows customers prefer charcoal broiling but the unit is more expensive. The manager's Minimum Attractive Rate of Return (MARR) is 20%. The manager estimates the costs and net benefits from the two alternatives as follows:

End of Year	Broiler	Grill
0	-\$2200	-\$1450
1-5	+990	+710

Use incremental rate of return analysis to determine which alternative the manager should choose.

****

Broiler RoR: $2{,}200 = 990(P/A, i\%, 5)$

$(P/A, i\%, 5) = \frac{2{,}200}{990} = 2.22$; $i = 35\% >$ MARR

Grill ROR: $1{,}450 = 710\,(P/A, i\%, 5)$

$(P/A, i\%, 5) = \frac{1{,}450}{710} = 2.04$; $i = 40\% >$ MARR

ΔRoR:

END OF YEAR	BROILER - GRILL
0	-750
1-5	+280

$750 = 280(P/A, i\%, 5)$

$(P/A, i\%, 5) = \frac{750}{280} = 2.679$; $i \cong 25\% >$ MARR

Choose higher cost alternative ∴ choose Broiler

** **6-10**

John Q. Customer has received his bill for the next 6 months premium on his auto insurance. The bill allows him two methods to pay his premium of $189.00. He can either pay the entire amount now, or he can pay $99.00 now, which is one half of the premium plus a $4.50 prepaid "service charge," and $94.50 in two months, the other half of his premium. The insurance company is, implicitly, offering John a "loan." What is the effective annual interest rate of the loan?

****

The net cash flows describing John's alternatives are as follows:

t	Pay Now	Half later	Later - Now
0	-189	-99	+90
1			
2		-94.50	-94.50

The effective monthly interest rate (rate of return) is:

$90 = 94.50(P/F, i_{mo}, 2) = 94.50(1 + i_{mo})^{-2}$

$i_{mo} = \sqrt{94.5/90} - 1 = 0.0247$

The effective annual interest rate is therefore:

$i = (1.0247)^{12} - 1 = 0.34$ or 34%

** **6-11**

Of two projects, A has the higher maintenance cost, but B has the higher investment cost. The incremental rate of return is 17.6%. State which alternative is preferred if the Minimum Attractive Rate of Return (MARR) is 20%.

****

If the MARR > Incremental Rate of Return, take the alternative with the lower investment cost. Therefore Alternative A is preferred.

6-12 ***

The table below summarizes incremental rate of return information for four mutually exclusive alternatives, one of which must be chosen. The table includes all possible comparisons of alternatives. For example, the value of 6% in column A represents the incremental rate of return at which alternatives A and B are equally attractive.

Alternative	First Cost	Overall Incremental Rate of Return	Incremental Rate of Return For Breakeven			
			A	B	C	D
A	$20,000	9%	--	6%	10%	10%
B	15,000	10	6%	--	14	12
C	10,000	8	10	14	--	10
D	5,000	6	10	12	10	--

(a) Make a table showing the range of Minimum Attractive Rate of Return (MARR) in which each alternative is preferred.

(b) If MARR = 7%, is the Net Present Worth of Alternative C positive, negative, or equal to zero?

****

a) Null alternative not available.

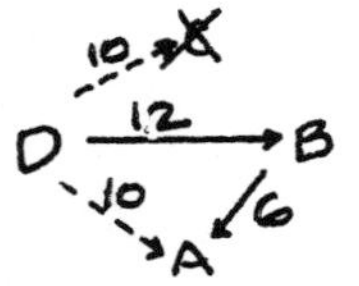

choose D if $12 < MARR$

B if $6 < MARR < 12$

A if $MARR < 6$

never choose C

b) Since ROR(8%) > MARR(7%) NPW > 0 (Positive)

6-13 ***

Is the following statement True or False?

If two projects each have a 15% Rate of Return, then the Incremental Rate of Return must also be 15%.

****

The statement is True.

** **6-14**

General Motors makes a "P-Car" automobile model. The standard P-Car is sold as a Chevy while a modified, more "luxurious" P-Car is sold as a Buick. Although the two cars are essentially the same, the Buick costs $15,000 while the Chevy costs $13,000. A friend of yours claims that the Buick is a better buy than the Chevy since the Buick has a higher resale value. The Buick may be sold after 4 years for $7500 while the Chevy can be sold for $5000.

(a) Calculate the incremental rate of return. Express your final answer as XX.XXX%.
(b) For what range of values of the Minimum Attractive Rate of Return (MARR), if any, is the Buick preferred?
(c) For what range of values of MARR, if any, is the Chevy preferred? (Ignore the "benefit"/"cost" of owning/operating a more luxurious car.)

****

a)

YEAR	BUICK	CHEVY	BUICK - CHEVY
0	-15,000	-13,000	-2,000
4	+7,500	+5,000	+2,500

$2{,}000 = 2{,}500\,(P/F, i\%, 4)$

$(P/F, i\%, 4) = .8000$

try $i = 5\%$	.8227	by interpolation:
$i = ?$	.8000	$i = 5\% + (6\% - 5\%)\left(\frac{.8227 - .8}{.8227 - .7921}\right)$
$i = 6\%$	.7921	

$i = 5.742\%$

b) MARR ≤ 5.74 select Buick

c) MARR > 5.74 select Chevy

** **6-15**

Would present worth analysis, using a given minimum attractive rate of return, give the same solution to a particular problem as the more difficult incremental or marginal analysis? Explain.

****

Yes. If the incremental rate of return in making a larger investment is less than the Minimum Attractive Rate of Return (MARR), the Net Present Worth of the larger investment will be less than for the lesser valued initial investment. Thus both methods select the same alternative.

6-16 **

A young engineering technology graduate is interested in buying a solar energy system for her home. She is investigating two particular systems shown below.

	Solarex	Soltech
Initial Cost	$10,000	$6,000
Annual Maintenance Cost	100/year	50/year
Annual Savings in Heating Costs	1,000/year	700/year
Salvage Value (end of life)	500	400
Expected Life	20 years	15 years

Use an analysis period of 20 years. Use SOYD depreciation for the second Soltech bought to determine its salvage value at the end of Year 20.

Calculate the incremental rate of return of the two systems and select the most desirable one based on a Minimum Attractive Rate of Return of 6%.

****

Year	Solarex (A)	Soltech (B)	A-B
0	-10,000	-6,000	-4,000
1-20	+ 900	+ 650	+ 250
Add 15	0	- 5,600	+ 5,600
Add 20	+ 500	+ 2,967	- 2,467

Using SOYD get Book Value of 2nd Soltech at end of year 20 (5 years after installed).

$$BV(SOYD) = 6{,}000 - \frac{5\left[15 - \frac{5-1}{2}\right]\left[6{,}000 - 400\right]}{\frac{15}{2}(15+1)} = \$2{,}967.$$

$NPW = 250(P/A, i\%, 20) + 5{,}600(P/F, i\%, 15) - 2{,}467(P/F, i\%, 20) - 4{,}000 = 0$

$i = 10\%$, $NPW = 250(8.514) + 5{,}600(.2394) - 2{,}467(.1486) - 4{,}000 = -897.45 < 0$ lower

$i = 5\%$, $NPW = 250(12.462) + 5{,}600(.4810) - 2{,}467(.3769) - 4{,}000 = 879.28 > 0$ higher

$i = 7\%$, $NPW = 250(10.594) + 5{,}600(.3624) - 2{,}467(.2584) - 4{,}000 = 40.46 > 0$ higher

$i = 8\%$, $NPW = 250(9.818) + 5{,}600(.3152) - 2{,}467(.2145) - 4{,}000 = -309.54 < 0$

interpolate: $\Delta ROR = 7\% + (1\%)\frac{(40.46 - 0)}{40.46 - (-309.54)} = \underline{7.12\%}$

$\Delta ROR = 7.12\% > MARR\ (6\%)$, choose higher cost, <u>Solarex</u>.

** **6-17**

A large company has the opportunity to select one of seven projects: A, B, C,...,G, or the null (Do Nothing) alternative. Each project requires a single initial investment as shown in the table below. Information on each alternative was fed into a computer program which calculated all the rates of return and all the pertinent incremental rates of return as shown in the table.

Project	Initial Investment	Incremental Rate of Return of Project Over						
		Null	A	B	D	C	E	F
A	$10,000	10%						
B	12,000	9	7%					
C	13,000	8	2	0.1%				
D	15,000	7	9	5%	9%			
E	16,000	6	5	1	6	3%		
F	18,000	5	8	2	5	5	5%	
G	23,000	7	3	8	7	4	3	2%

For example, the rate of return for Project A is 10% and the incremental rate of Project C minus Project B (i.e., C–B) is 0.1%.

(a) Determine the range of values of the Minimum Attractive Rate of Return (MARR) for which the null (do nothing) alternative is the preferred alternative. Determine the range of values of the MARR for which Project A is preferred, for which B is preferred, and so on. (Note that there may be no range of values of the MARR for which certain projects are preferred.)

(b) Now suppose that for reasons beyond your control, Project A is not longer an available alternative. Now repeat (a) for this situation.

****

a)

If:	Select:
MARR > 10%	NULL
10% ≥ MARR > 9%	A
9% ≥ MARR > 5%	D
5% ≥ MARR > 2%	F
2% ≥ MARR	G

b)

If:	Select:
MARR > 9%	NULL
9% ≥ MARR > 8%	B
8% ≥ MARR	G

6-18 **

	A	B
Investment Cost	$20,000	$10,000
Annual Maintenance Cost	2,000	4,000
Salvage	5,000	1,000
Useful Life	20 years	20 years

(a) Determine the incremental rate of return between the two projects, to the nearest 1/100 percent.

(b) Which project should be chosen if MARR = 18%? (No additional calculations needed.)

****

a) $NPW_H - NPW_L = [-20{,}000 - 2{,}000(P/A, i\%, 20) + 5{,}000(P/F, i\%, 20)]$
$- [10{,}000 - 4{,}000(P/A, i\%, 20) + 1{,}000(P/F, i\%, 20)] = 0$
$= -10{,}000 + 2{,}000(P/A, i\%, 20) + 4{,}000(P/F, i\%, 20) = 0$

Try $i = 20\%^*$: $-10{,}000 + 2{,}000(4.870) + 4{,}000(.0261) = -155.60$

$i = 18\%$: $-10{,}000 + 2{,}000(5.353) + 4{,}000(.0365) = 852.00$

$\Delta ROR = 18 + 2\left(\frac{852}{852 + 155.6}\right) = \underline{19.69\%}$

b) $\Delta ROR > MARR$, choose higher cost project, <u>choose A</u>

* To find a trial i: $(P/A, i, 20) \approx \frac{10{,}000}{2{,}000} = 5$
From tables $i \approx 20\%$

6-19 **

Project A ($10,000 investment) has a rate of return of 25%. Project B ($20,000 investment) has a rate of return of 15%. Which of the following is true?

(a) 25% < Incremental Rate of Return
(b) 15% < Incremental Rate of Return < 25%
(c) Incremental Rate of Return < 15%

****

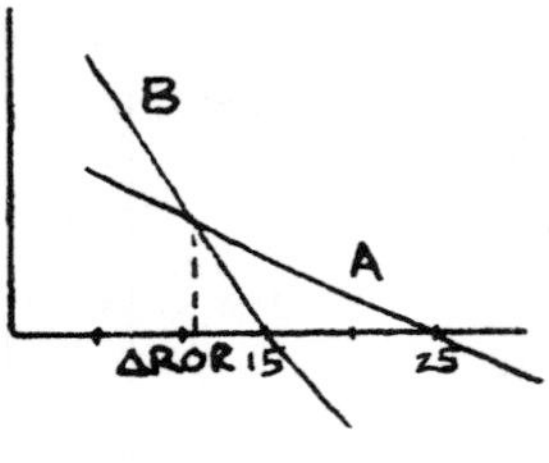

Higher investment project has steeper sloped line. <u>Choose C.</u>

** **6-20**

A store owner is considering installing a security system to reduce shoplifting losses. Without the system, the owner figures 100 shoplifters each year are not caught and that next year each shoplifter will steal, on the average, $6.45 of merchandise. This average will increase at the rate of 35 cents each year thereafter because the owner plans to increase his prices at that rate to keep up with inflation.

With the system, the owner estimates half the shoplifters will not attempt it. The other half wil be caught in the attempt, and because of the apparent increase in the risk of being caught, they will try to steal higher priced goods. The owner estimates during the first year of operations, each shoplifter will be caught with, on the average, $10.96 of merchandise. This average will also increase at 35 cents per year. The system costs $1,195.60 installed and has a life of 5 years. What is the rate of return on this investment?

****

The relevant cash flows are:

t	without system	with system	difference (with - without)
0	0	-1,195.60	-1,195.60
1	-100(6.45)		645
2	-100(6.80)		680
3	-100(7.15)		715
4	-100(7.50)		750
5	-100(7.85)		785

Notice that with the system the fact that half of the shoplifters will attempt to steal higher priced goods is irrelevant since they will be caught and the merchandise returned to the store. The other half don't attempt it and thus there are no other costs after installation. The rate of return, i, is:

$NPV = 0 = -1{,}195.60 + 645(P/A, i\%, 5) + 35(P/G, i\%, 5)$

$i = 50\%$

6-21 **

A large food corporation is considering the development and production of four types of beverages, ranging from juices to instant drinks. The type of markets, margins of profit, sales volume and technology needed are quite different in each case. The following summarizes the economic aspects of the projects.

	Product 1	Product 2	Product 3	Product 4
Equipment costs	\$597,500	\$446,100	\$435,700	\$249,800
Installation costs	250,000	150,000	200,000	100,000
Expected annual profits	212,000	145,000	168,000	100,000

This company's MARR is 13%. Assume the projects' lives will be 6 years. Which alternative is the most profitable?

****

1. Calculate the individual ROR:

PW Costs = PW Ben

$P = A(P/A, ROR, 6) \Rightarrow (P/A, ROR, 6) = P/A$

Product 1: $(P/A, ROR_1, 6) = \frac{597,500 + 250,000}{212,000} = 3.9976$

$ROR_1 \approx 13\%$ ∴ attractive

Product 2: $(P/A, ROR_2, 6) = \frac{446,100 + 150,000}{145,000} = 4.111$

$ROR_2 \approx 12\%$ ∴ not attractive

Product 3: $(P/A, ROR_3, 6) = \frac{435,700 + 200,000}{168,000} = 3.7839$

$ROR_3 \approx 15\%$ ∴ attractive

Product 4: $(P/A, ROR_4, 6) = \frac{249,800 + 100,000}{100,000} = 3.498$

$ROR_4 \approx 18\%$ ∴ attractive

2. Incremental Analysis:

$PW\ Cost_4 < PW\ cost_3 < PW\ Cost_1$

a) Increment 3-4:

$\Delta cost = (435,700 + 200,000) - (249,800 + 100,000) = 285,900$

$\Delta Ann\ Ben = 168,000 - 100,000 = 68,000$

$PW_{\Delta cost} = PW_{\Delta Ben} \Rightarrow 285,900 = 68,000(P/A, \Delta ROR_{34}, 6)$

$(P/A, \Delta ROR_{34}, 6) = 4.2044 \Rightarrow \Delta ROR_{34} \approx 11.2\% < MARR$

∴ Keep 4

b) Increment 1-4:

$\Delta Cost = (597,500 + 250,000) - (249,800 + 100,000) = 497,700$

$\Delta Ann\ Ben. = 212,000 - 100,000 = 112,000$

$(P/A, \Delta ROR_{14}, 6) = \frac{497{,}700}{112{,}000} = 4.4437$

$\Delta ROR_{14} \approx 9.3\% << MARR \Rightarrow \therefore$ keep 4

Product 4 is the most profitable.

*** **6-22**

Consider four mutually exclusive alternatives that have 10-year useful lives and no salvage value. If the Minimum Attractive Rate of Return (MARR) is 6%, which alternative should be selected?

	A	B	C	D
Initial Cost	$3000	$2000	$5000	$4000
Uniform Annual Benefit	447	259	885	651

****

Alt. A : $(P/A, i, 10) = 3{,}000/447 = 6.711$ $i = 8\%$

Alt. B : $(P/A, i, 10) = 2{,}000/259 = 7.722$ $i = 5\%$

Alt. C : $(P/A, i, 10) = 5{,}000/885 = 5.650$ $i = 12\%$

Alt. D : $(P/A, i, 10) = 4{,}000/651 = 6.144$ $i = 10\%$

Reject alternative B because ROR < 6%. Organize the other alternatives in increasing order of cost for incremental analysis:

	D−A	C−A
Δ Cost	1,000	1,000
Δ UAB	204	234
Δ ROR	15.7%	19.5%

select alternative C largest cost

$(P/A, i, 10) = 1{,}000/204 = 4.902$

15% ⇒ 5.019

x ⇒ 4.902 $x = 3\left(\frac{.117}{.525}\right) = .67 \rightarrow i = 15.67\%$

18% ⇒ 4.494

$(P/A, i, 10) = 1{,}000/234 = 4.274$

18% ⇒ 4.494

x ⇒ 4.274 $x = 2\left(\frac{.22}{.302}\right) = 1.46 \rightarrow i = 19.6\%$

20% ⇒ 4.192

Select alternative with largest cost, select alternative C

6-23 **

A local brewing company is considering the marketing of one of its main brands in new non-returnable bottles. This brand is currently being marketed in returnable bottles, and will continue to do so. The new presentation is expected to increase profits (due to increase in sales, and to a higher margin of profit per bottle) in the next 10 years, in the amounts given below. This will require, however, an investment of $165,000 for packaging equipment, with no salvage value at the end of the ten years. If the company's MARR is 12%, should they invest in the project?

Year	Additional Volume* of returnable bottles	Additional Volume* of new non-returnable bottles
1	-3000 units	+4000 units
2	-2500	+4000
3	-2000	+4000
4	-1500	+4000
5	-1000	+4000
6	-500	+4000
7	0	+4000
8	+500	+4000
9	+1000	+4000
10	+1500	+4000

* Compared to Year 0.

The profit is $5.15 per unit for returnable bottles and $8.55 for non-returnable bottles. Negative values refer to a decrease in sales. This decrease is projected due to some of the current consumers changing to the new non-returnable bottle.

****

Calculate first the total profit if the new bottle is introduced:

YEAR	Additional Profit: Returnable Bottles	Additional Profit: non-returnable	Additional Total Profit
1	-15,450*	34,200**	18,750
2	-12,875	34,200	21,325
3	-10,300	34,200	23,900
4	- 7,725	34,200	26,475
5	- 5,150	34,200	29,050
6	- 2,575	34,200	31,625
7	0	34,200	34,200
8	2,575	34,200	36,775
9	5,150	34,200	39,350
10	7,725	34,200	41,925

* from 3,000 units × 5.15 $/unit = $15,450
** from 4,000 units × 8.55 $/unit = $34,200
The cash flow, therefore, is as follows:

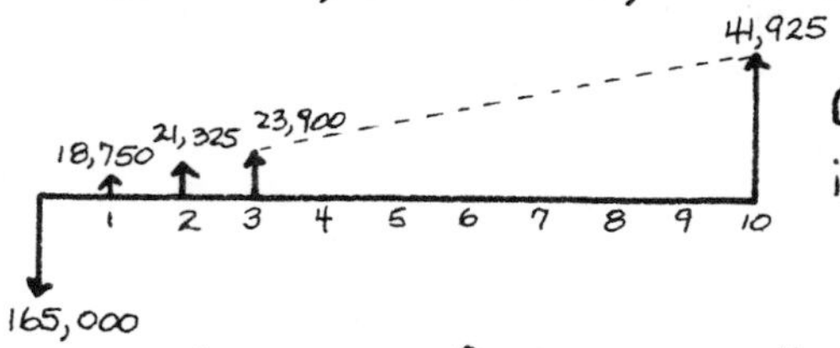

(note: there is a constant increase of $2,575 per year)

Therefore, the NPW of profits for the next 10 yrs. is:

$NPW = 0 = 18{,}750(P/A, i, 10) + 2{,}575(P/G, i, 10) - 165{,}000$

for $i = 10\%$ $NPW = 18{,}750(6.145) + 2{,}575(22.891) - 165K = 9163.07$

for $i = 12\%$ $NPW = 18{,}750(5.65) + 2{,}575(20.254) - 165K = {}^{-}6{,}908.45$

then: $i \approx 10\% + 2\% \left(\dfrac{9163.07 - 0}{9163.07 + 6908.45}\right) = 11.14\%$

since $i <$ MARR (11.14% < 12%) ⟹ Not profitable to invest in the new non-returnable bottle.

** **6-24**

The following equation describes the differences between two projects.

$$NPW_H - NPW_L = 3000(P/A,i,10) + 4000(P/F,i,10) - 10{,}000$$

(a) Find the incremental rate of return to 1/100 of a percent.
(b) Which alternative should be chosen if MARR = 10%? (No additional calculations are needed.)

****

a)

Trial i: $(P/A, i\%, 10) \approx \dfrac{10{,}000}{3{,}000} = 3.33$ ignoring salvage

$i \approx 25\%$

Try $i = 25\%$: $3{,}000(3.571) + 4{,}000(.1074) - 10{,}000 = +1{,}142.6$

Try $i = 30\%$: $3{,}000(3.092) + 4{,}000(.0725) - 10{,}000 = -434.$

$\Delta RoR = 25 + 5\left(\dfrac{1{,}142.6}{1{,}142.6 + 434}\right) = \underline{28.62\%}$

b)

The higher cost project since ΔRoR > MARR

6-25 **

Many persons believe that Net Present Value (often called Net Present Worth) and Rate of Return can lead to a contradictory choice in a decision between alternatives. The following example demonstrates that a proper incremental analysis between alternatives X and Y would result in the same decision. The Minimum Attractive Rate of Return is 15%.

Year	X	Y	Y-X	
0	-$10,000	-$20,000	-$10,000	
1	+5,500	0	-5,500	
2	+5,500	0	-5,500	
3	+5,500	+40,000	+34,500	
Rate of Return:	30%	26%	24.1%	>15%
Net Present Value:	$2,558	$6,300	$3,742	>0

When shown an example such as this, some people still remain skeptical. Among the remarks they frequently make is the following:

"I understand what you're saying about an incremental analysis, nonetheless it seems to me that if I select Y, as you suggest, with a rate of return of 26%, I'll forego the opportunity to invest at 30% and that doesn't make sense to me. Why should I give up the chance to invest at 30% to invest at a lower rate of return?"

Respond to this remark. By selecting Y is the investor foregoing the opportunity to invest at 30%? Be specific and concise.

****

By selecting Y, one does not forego the opportunity to invest at 30%. This can be more easily seen by writing the cash flows as follows:

	ALTERNATIVE X		ALTERNATIVE Y	
YEAR	X + CASH LEFTOVER		X + (Y-X) = Y	
0	-10,000	-10,000	-10,000	-10,000
1	+5,500		+5,500	-5,500
2	+5,500		+5,500	-5,500
3	+5,500		+5,500	+34,500
Rate of Return	30%	MARR=15%	30%	24.1%

If X is selected $10,000 will be invested at 30% and $10,000 at (presumably) the MARR of 15%. Notice that Y can be divided into two parts, one part has the same cash flows as X and the other part has the same cash flows as Y-X. Thus, if Y is selected, $10,000 will be invested at 30% and the difference 24.1%. Regardless of whether X or Y is selected $10,000 will be invested at 30%, and the difference will be invested at either 15% or 24.1% depending on the choice.

** **6-26**

A farmer needs to purchase a new grain combine to harvest his rice. He is considering two alternatives, a K-E combine and a J-I combine. The K-E combine costs $100,000 and is espected to produce revenues of $122,000 per year. The J-I machine costs $135,000 but is expected to produce revenues of $128,000 per year because of its higher efficiency.

Operating cost for both machines is $100,000 per year. Life of each machine is 7 years. If the farmer's Minimum Attractive Rate of Return (MARR) is 8%, which one should he buy? Use Incremental Rate of Return Analysis.

****

Assume no salvage

	K-E	J-I	J-I − K-E
First Cost	$100,000	$135,000	35,000
Ann. Bene.	122,000	128,000	6,000
Ann. Cost	100,000	100,000	0

Compute ROR for each alternative:

K-E: $NPW = 0 = -100{,}000 + 22{,}000\,(P/A, i\%, 7)$

$(P/A, i\%, 7) = 4.55$

$i = ROR \cong 12\% > MARR$

J-I: $NPW = 0 = -135{,}000 + 28{,}000\,(P/A, i\%, 7)$

$(P/A, i\%, 7) = 4.82$

$i = ROR \cong 10\% > MARR$

ΔROR: $NPW = -35{,}000 + 6{,}000\,(P/A, i\%, 7)$

$(P/A, i\%, 7) = 5.83$

$i = \Delta ROR \cong 4.75\% < MARR$

choose lower cost combine : K-E

6-27 **

A firm needs to increase its manufacturing floor space to at least 125,000 square feet. Seven independent plant sites have been identified in existing buildings in seven cities. The floor space, the first cost to renovate the plant and the uniform annual revenue for each location is shown below. The life of each plant is 8 years and the Minimum Attractive Rate of Return is 10% for this firm. Also shown below are the rates of return for each location. The firm will consider any combination of locations that satisfies its floor space requirements.

Location	Floor Space	First Cost	Uniform Annual Revenue	Rate of Return
Denver	75,000 ft^2	$300,000	$ 75,000	18.6%
Dallas	125,000	750,000	187,000	18.5
San Antonio	50,000	450,000	117,000	19.9
Los Angeles	125,000	550,000	122,000	14.9
Cleveland	50,000	150,000	25,000	6.9
Atlanta	75,000	200,000	49,000	18.0
Chicago	100,000	100,000	20,000	11.8

(a) If the firm has no budget constraint, what course of action would you recommend?

(b) If the firm has a limited budget and cannot spend more than $600,000, what course of action would you recommend?

****

From the rates of return we can eliminate Cleveland since the 6.9% rate of return is less than the 10% MARR.

We can use any measure of worth to evaluate the remaining locations. This solution uses Net Present Value (NPV), although Equivalent Uniform Cash Flow would also be a good choice in this problem because the revenues are already uniform.

The combinations of plant sites that satisfy the space requirements are shown below. Also shown are the NPV's for each plant and for each combination.

Plant	Combination 1	2	3	4	5	6	Plant NPV
Den.	75				75		*100.12
Dal.		125					247.63
S.A.	50		50			50	174.19
L.A.				125			100.86
Atl.			75		75		61.41
Chi.						100	6.70
ft^2	125	125	125	125	150	150	

Plant	1	2	3	4	5	6	Plant NPV
Comb. NPV	274.31	247.63	235.60	100.86	161.53	180.89	
Total F.C.	750	750	650	550	500	550	

a) If funds are unlimited, the Denver and San Antonio sites are the most economical.

b) If funds are limited to $600,000, combinations 1 to 3 are eliminated, and the best of the remainder is San Antonio and Chicago.

** **6-28**

The table below summarizes incremental rate of return information for four possible mutually exclusive alernatives.

Alternative	First Cost	Overall Rate of Return	Incremental Rate of Return for Breakeven A	B	C	D
A	$18,000	9%	--	9%	12%	11%
B	12,000	9	9%	--	6%	15%
C	15,000	8	12%	6%	--	10%
D	10,000	6	11%	15%	10%	--

The table includes all posible comparisons of alternatives. For example, the value of 9% in column A represents the Incremental Rate of Return at which Alternatives A and B are equally attractive.

(a) Make a choice table showing the range of Minimum Attractive Rate of Return (MARR) in which each alternative is preferred.

(b) If MARR = 9%, is the Net Present Worth of Alternative B positive, negative, or zero?

****

(a) The null (do nothing alternative) is available since overall ROR is given and nothing expressly forbids its choice.

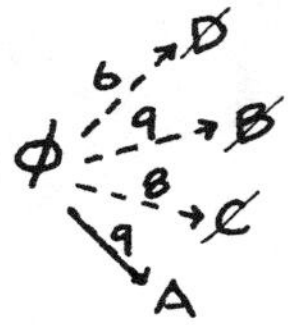

choose NULL if 9% < MARR
choose A if MARR < 9%
never choose B, C or D

(b) It is zero.

6-29 **

Consider five mutually exclusive alternatives that have 10-year useful lives and no salvage value. If the Minimum Attractive Rate of Return (MARR) is 9%, which alternative should be selected?

	A	B	C	D	E
Initial cost	$4000	$5000	$2000	$3000	$6000
Uniform Annual Benefit	651	885	259	447	1195

****

Determine individual rate of return:

Alt. A: $(P/A, i, 10) = 4000/651 = 6.144$ $i = 10\%$

Alt. B: $(P/A, i, 10) = 5,000/885 = 5.650$ $i = 12\%$

Alt. C: $(P/A, i, 10) = 2,000/259 = 7.722$ $i = 5\%$

Alt. D: $(P/A, i, 10) = 3,000/447 = 6.711$ $i = 8\%$

Alt. E: $(P/A, i, 10) = 6,000/1,195 = 5.020$ $i = 15\%$

reject alternatives C and D because these rates of return are less than the minimum attractive rate of return.

incremental analysis:

	B−A	E−B
Δ Cost	1,000	1,000
Δ UAB	234	310
Δ ROR	19.5%	28.6%

choose E, largest cost alternative

(B−A): $(P/A, i, 10) = 1,000/234 = 4.2735$

18 ⇒ 4.494

x ⇒ 4.2735 $\frac{x}{2} = \frac{.220}{.302}$ → $x = 1.456$

20 ⇒ 4.192 $i = 19.5\%$

(E−B): $(P/A, i, 10) = 1,000/310 = 3.2258$

25 ⇒ 3.571 $\frac{x}{5} = \frac{.345}{.479}$ → $x = 3.60$

x ⇒ 3.226 $i = 28.6\%$

30 ⇒ 3.092

** **6-30**

The table below summarizes incremental rate of return information for five mutually exclusive alternatives, one of which must be chosen (that is, the null or do nothing alternative is not available). The table includes all possible comparisons of alternatives. For example, the value of 22% in column A represents the incremental rate of return at which alternatives A and C are equally attractive.

			Incremental Rate of Return For Breakeven				
Alternative	First Cost	Overall Rate of Return	A	B	C	D	E
A	$80,000	19%	--	29%	22%	25%	24%
B	60,000	16	29	--	14	23	21
C	40,000	17	22	14	--	35	29
D	30,000	12	25	23	35	--	17
E	20,000	8	24	21	29	17	--

The questions below use this information to test whether you understand how to use incremental rate of return methods without the need for you to make extensive calculations.

(a) If the Minimum Attractive Rate of Return (MARR) = 35%, which alternative is best?
(b) In what range of MARR is alternative C the best choice?
(c) In what range of MARR is alternative B the best choice?
(d) If the null were to become available, which alternatives would never be chosen?

****

a)

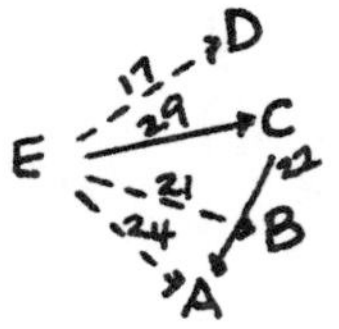

E is the best alternative

b) 22% < MARR < 29%

c) None

d)

Alternatives B, C, D, E will never be chosen

If MARR > 19% choose ϕ
If MARR < 19% choose A

6-31 ***

A manufacturer, whose Minimum Attractive Rate of Return (MARR) is 20%, is considering two alternative automated material handling systems, A and B, to replace his current manually operated system. The cash flows that describe each automated system with respect to the current system are given below. Using rate of return as the decision criterion, what course of action do you recommend?

Year	System A	System B
0	-$10,000	-$11,500
1-15	+1,029	+1,780

****

The computed rate of return for System A is 6% and for System B it is 13%. Since both rates of return are less that the minimum required 20%, neither system is more economical than the current system.

Notice the incremental rate of return of B-A:

Year	System A	System B	System B - System A
0	-$10,000	-$11,500	-$1,500
1-15	+1,029	+1,780	+751

-1,500 = 751(P/A,i%,15) Solving for i we find the incremental rate of return of B-A = 50%, but this figure is irrelevant since neither system alone, compared to the current system, is economical. Retain the manually operated system.

7 OTHER ANALYSIS TECHNIQUES

FUTURE WORTH

** **7-1**

A mortgage of $20,000 for 30 years, with monthly payments at 10% interest is contemplated. At the last moment you receive news of a $5000 gift from your parents to be applied to the principal. Leaving the monthly payments the same, what amount of time will now be required to pay off the mortgage and what is the amount of the last payment (assume any residual partial payment amount is added to the last payment)?

****

$A = 20{,}000(A/P, \frac{10\%}{12}, 360)$

$= 20{,}000(.0087757) = \$175.51$ monthly payment

After reduction of P to 15,000

$15{,}000 = 175.51(P/A, \frac{10\%}{12}, n)$; $P/A = 85.463$

Try n = 12 years; $P/A = 83.676$

Try n = 13 years; $P/A = 87.120$

Interpolate, <u>n = 12.52 years</u>

at 12 years 6 months: $P = 175.51(P/A, 0.833\%, 150)$

$= 175.51(85.441) = 14{,}995.75$

Residual = 4.25

Last Payment = Value of residual at time of last payment + last payment

Last Payment $= 4.25(F/P, \frac{10\%}{12}, 150) + 175.51$

$= 4.25(3.4706) + 175.51 = 190.26$

7-2 **

A woman deposited $10,000 into an account at her credit union. The money was left on deposit for 10 years. During the first five years the woman earned 15% interest (nominal), compounded monthly. The credit union then changed it's interest policy so that the second five years the woman earned 18% interest (nominal), compounded quarterly.

(a) How much money was in the account at the end of the 10 years?
(b) Calculate the rate of return that the woman received.

****

part a:

at the end of 5 years: [1]

$F = 10{,}000(F/P, 1\frac{1}{4}\%, 60) = 10{,}000(2.107) = 21{,}070$

at the end of 10 years: [2]

$F = 21{,}070(F/P, 4\frac{1}{2}\%, 20) = 21{,}070(2.412) = \underline{50{,}820.84}$

part b:

$10{,}000(F/P, i, 10) = 50{,}820.84$

$(F/P, i, 10) = 5.082084$

try $i = 15\%$ $(F/P, i, 10) = 4.046$

try $i = 18\%$ $(F/P, i, 10) = 5.234$

interpolate: $i = 15\% + (18\% - 15\%)\left(\frac{4.046 - 5.082}{4.046 - 5.234}\right)$

$\underline{i = 17.616\%}$

1 - $i\% = \frac{15\%}{12} = 1\frac{1}{4}\%$; (5 years)(12 per/yr) = 60

2 - $i\% = \frac{18\%}{4} = 4\frac{1}{2}\%$; (5 years)(4 per/yr) = 20

7-3 **

A woman deposited $100 per month in her savings account for 24 months at 6% interest, compounded monthly. Then for five years she made no deposits or withdrawals. How much is the account worth after seven years?

****

Assume monthly compounding

$FW = 100(F/A, \frac{1}{2}\%, 24)(F/P, \frac{1}{2}\%, 60)$

$= 100(25.432)(1.349) = \underline{\$3{,}430.78}$

** **7-4**

On January 1st a sum of $1200 is deposited into a bank account that pays 12% interest, compounded monthly. On the first day of each succeeding month $100 less is deposited (so $1100 is deposited February 1st, $1000 on March 1st, and so on). What is the account balance immediately after the December 1st deposit is made?

****

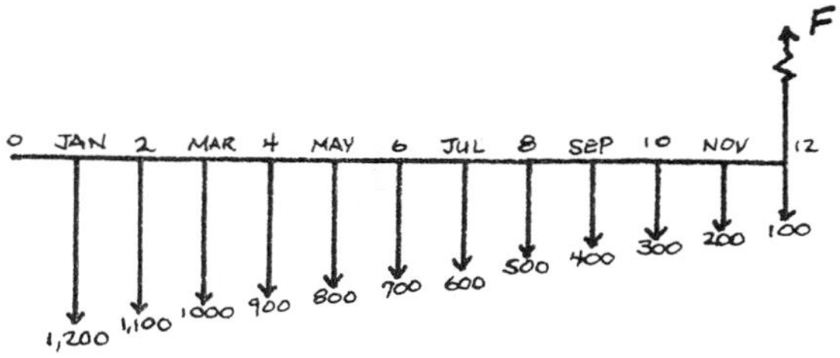

$$P = 1{,}200(P/A, 1\%, 12) - 100(P/G, 1\%, 12)$$
$$= 1{,}200(11.255) - 100(60.569) = 7{,}449.1$$
$$F = 7{,}449.1\,(F/P, 1\%, 12) = 7{,}449.1(1.127) = \underline{8395.14}$$

** **7-5**

A 25-year-old engineer named Milton begins working for a salary of $30,000 per year when he graduates from college. From his first monthly paycheck, he notices 7% of his salary is deducted and paid into Social Security, and his employer pays a like amount. In effect, Milton finds that 14% of his salary is being taken by the government for this mandatory program.

Assuming that Milton's contribution to Social Security is 14% of $30,000 = $4200 per year, and that Milton works for the same salary until he is 65, how much will he have effectively contributed into the Social Security program. Assume a 10% interest rate.

****

$$F = A(F/A, i, n) \quad \text{where } A = \$4{,}200;\ i = 10\%;\ n = 40\text{ yrs.}$$
$$= 4{,}200(442.593) = \underline{1{,}858{,}890.60}$$
$$F = A\left[\frac{(1+i)^n - 1}{i}\right] = 4{,}200\left[\frac{1.10^{40} - 1}{0.10}\right] = \underline{1{,}858{,}888.73}$$

7-6 **

An engineer is considering the purchase of a new set of batteries for a tractor. Given the cost, annual benefit and useful life, conduct a Net Future Worth (NFW) analysis to decide which alternative to purchase if i = 12%.

	A	B
Cost	$150	$90
Annual benefit	40	40
Useful life	6 yrs	3 yrs

****

Alternative A:

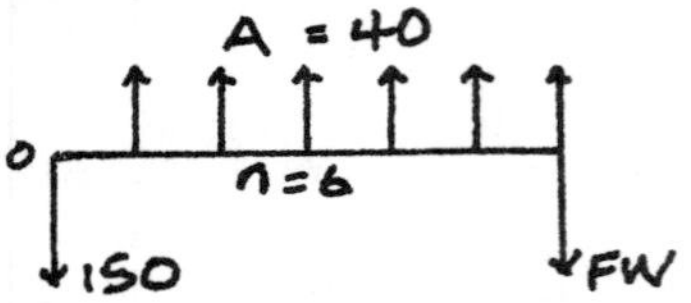

NFW = 40(F/A, 12%, 6) − 150(F/P, 12%, 6) = 40(8.115) − 150(1.974)

= 324.6 − 296.1 = +28.5

Alternative B:

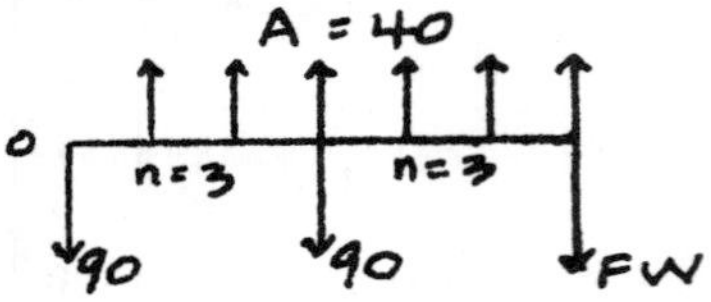

NFW = 40(F/A, 12%, 6) − 90(F/P, 12%, 6) − 90(F/P, 12%, 3)

= 40(8.115) − 90(1.974) − 90(1.405)

= 324.6 − 177.66 − 126.45 = +20.49

choose Alternative A, largest NFW

7-7 **

How much money would be in an account if $1000 is deposited in a bank at 12% interest, compounded semiannually, for 3 years?

****

$i = 12/2 = 6\%$; $n = 2 \times 3 = 6$ periods

$F = P(1+i)^n = \$1{,}000(1.06)^6 = \$1{,}000(1.41852) = \underline{\$1418.52}$

** **7-8**

To start a programming business, a computer science student bought a personal computer for $5000 to be paid for with a down payment of $1000 and 24 monthly payments at 18%, compounded monthly.

(a) How much are the payments?

(b) How much of the first payment is interest?

(c) The bank is terminating the loan now, six months after the computer was bought, because the fourth and fifth payments had not been made (the first three monthly payments were made). How much money must he get from his parents (or another source) to pay off the loan balance in full now to prevent losing the computer? Assume the only "penalty" is that interest continues to compound on any unpaid balance.

****

(a) Price = $5,000
Downpayment = 1,000
Loan 4,000
$i = \frac{18}{12} = 1\frac{1}{2}\%$

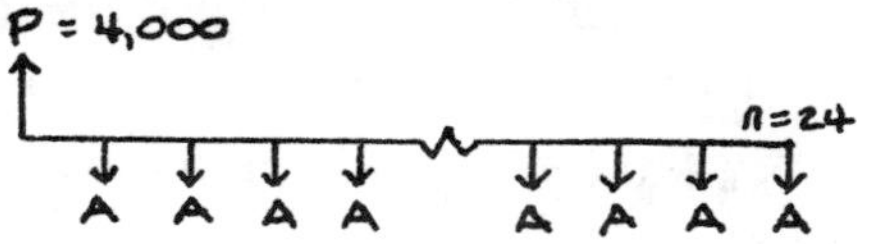

$A = 4{,}000\,(A/P, 1\frac{1}{2}\%, 24)$
$= 4{,}000\,(.0499) = \$199.60$

(b) Interest $= Pi = 4{,}000\,(.015) = \$60.$

(c) Amount owed equals the equivalent of unpaid payments determined at the time of concern (now).

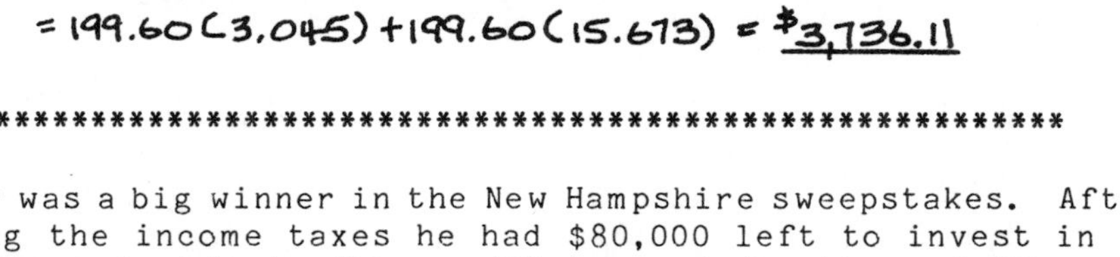

$X = 199.60\,(F/A, 1\frac{1}{2}\%, 3) + 199.60\,(P/A, 1\frac{1}{2}\%, 18)$
$= 199.60\,(3.045) + 199.60\,(15.673) = \$3{,}736.11$

** **7-9**

Harry was a big winner in the New Hampshire sweepstakes. After paying the income taxes he had $80,000 left to invest in an investment fund that will pay 10% interest for the next 20 years. How much money will Harry receive at the end of the 20 years?

****

$F = P(F/P, 10\%, 20) = 80{,}000(6.727) = \$538{,}160.00$

7-10 **

David has received $20,000 and wants to invest it for 12 years. There are three plans available to him.

(a) A Savings Account. It pays 5-3/4% per year, compounded daily.
(b) A Money Market Certificate. It pays 8-3/4% per year, compounded semiannually.
(c) An Investment Account. Based on past experience it is likely to pay 11.5% per year.

If David does not withdraw the interest, how much will be in each of the three investment plans at the end of 12 years?

****

a) $FW = F = P(1+i)^n$

effective interest rate : $(1+i)^m - 1 = EIR$

$EIR = (1 + \frac{.0575}{365})^{365} - 1 = (1+.00016)^{365} - 1 = .05918$

$\therefore FW = \$20{,}000(1+.05918)^{12} = \$39{,}872.14$

b) $FW = F = P(1+i)^n$

effective interest rate : $(1+i)^m - 1 = EIR$

$EIR = (1 + \frac{.0875}{2})^2 - 1 = (1+.04375)^2 - 1 = .089414$

$\therefore FW = \$20{,}000(1+.089414)^{12} = \$55{,}891.49$

c) $FW = P(1+i)^n$

$= 20{,}000(1+0.1150)^{12} = \$73{,}846.24$

Choose plan C since this plan yields the highest return at the end of 12 years.

7-11 **

How long will it take for $300 to triple at a 5% per year interest rate?

****

$F = 3P = P(F/P, 5\%, n) \Rightarrow (F/P, 5\%, n) = F/P = 3$

$3 = (F/P, 5\%, n)$

n	(F/P, 5%, n)
22	2.92526
≈ 22.5 ←	3.000
23	3.07152

$\rightarrow (F/P, 5\%, 22.5) = 3 \Rightarrow$ 22.5 years required to triple

** **7-12**

An annuity is established by the payment of $150 per month for eight years with interest to be calculated at 7-1/2%. The company retains these funds in an account from which they propose to pay you $1530 per year for life (an actuarial period of 18 years for your age). If interest is assumed to continue at 7-1/2%, what is the lump-sum profit to the company at the end of the pay-in period? Assume monthly compounding for the payment period.

****

Assume monthly compounding

$FW = 150(F/A, .075/12, 96) = 150(130.995) = 19,649$

$i_{eff} = (1 + .075/12)^{12} - 1 = .0776$

PW of payment $= 1,530(P/A, 7.76\%, 18)$

$= 1,530(9.530) = 14,581$

Profit at end of pay-in period $= 19,649 - 14,581 = \underline{\$5,068}$

** **7-13**

A person would like to retire 10 years from now. He currently has $32,000 in savings, and he plans to deposit $300 per month, starting next month, in a special retirement plan. The $32,000 is earning 10% interest, while the monthly deposits will pay him 9% nominal annual interest. Once he retires, he will collect the two sums of money, and being conservative in his calculations, he expects to get a 6% annual interest rate after year 10. Assuming he will only spend the interest he earns, how much will he collect in annual interest, starting in year 11?

****

a) Savings: $F = 32,000(F/P, 10\%, 10) = 32,000(2.594) = 83,008$

b) Monthly deposits: $F = 300(F/A, \frac{9}{12}\%, 120) = 300(193.514) = 58,054.2$

The total amount to deposit at the end of year 10 is:

$F_T = 83,008 + 58,054.2 = 141,062.2$

The interest to collect per year $= 141,062.2 \times 0.06 = 8,463.73$

7-14 ***

Mary wants to accumulate a sum of $20,000 over a period of 10 years to use as a downpayment for a house. She has found a bank that pays 12% interest compounded monthly. How much must she deposit four times a year (once each three months) to accumulate the $20,000 in ten years?

****

In this problem Mary's deposits do not match the interest period. One solution is to compute what her monthly deposit would need to be, and then the equivalent deposit each 3 months.

Monthly deposit (A) = $F(A/F, i\%, n) = 20{,}000(A/F, 1\%, 120)$

$= 20{,}000(0.0043) = \$86$

Equivalent deposit at end of each 3-months:

$= A(F/A, 1\%, 3) = 86(3.030) = \$261$

Note that the compound interest table only provided two digit accuracy.

<u>Hand Calculator Solution</u>:

monthly deposit (A) $= 20{,}000\left[\frac{0.01}{(1+0.01)^{120}-1}\right] = 86.942$

equivalent deposit at end of each 3-months:

$= 86.942\left[\frac{(1+0.01)^{3}-1}{0.01}\right] = \$263.44$

7-15 ***

If $5000 is deposited into a savings account that pays 8% interest, compounded quarterly, what will the balance be after 6 years? What is the effective interest rate?

****

$F = P(1+i)^n$

$= 5{,}000(1.02)^{24} = 5{,}000(1.6084) = \underline{8{,}042.19}$

$i_e = (1 + r/m)^m - 1$

$= (1 + .08/4)^4 - 1 = .08243 = \underline{8.243\% \text{ per yr.}}$

** **7-16**

Six months ago a local resident bought a new car for which he borrowed $4000 to be repaid over 24 months at 18% compounded monthly. He has just won the lottery and decides to completely pay off the car at the time of the seventh payment. How much is this final payment?

****

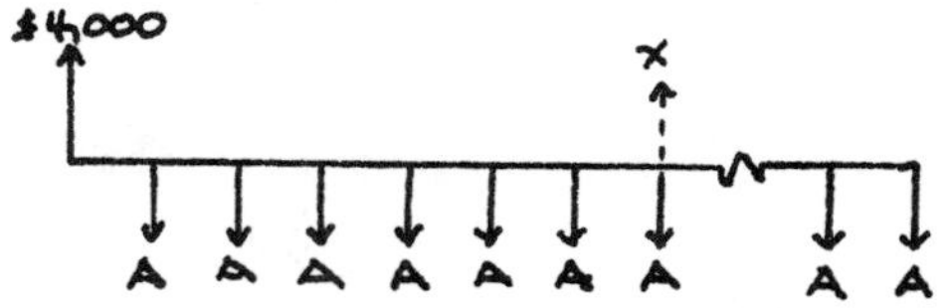

$i = 18\%/12 = 1\frac{1}{2}\%$

$A = 4{,}000(A/P, 1\frac{1}{2}\%, 24) = 4{,}000(.0499) = \$199.60$

$X = \$199.60 + 199.60(P/A, 1\frac{1}{2}\%, 17) = 199.60 + 199.60(14.908)$

$= \underline{\$3{,}175.24}$

** **7-17**

Starting now, deposits of $100 are made each year into a savings account paying 6% compounded quarterly. What will be the balance immediately after the deposit made 30 years from now?

****

$i = 6\%/4 = 1\frac{1}{2}\%$

$F = 100(A/F, 1\frac{1}{2}\%, 4)(F/A, 1\frac{1}{2}\%, 120) + 100(F/P, 1\frac{1}{2}\%, 120)$

$= 100(.2444)(331.288) + 100(5.969) = \underline{\$8{,}693.58}$

** **7-18**

You invest $1000 in a bank at 8 percent nominal interest. Interest is continuously compounded. What will your investment be worth

(a) at the end of 1 year?
(b) at the end of 36 months?

****

a) $F = Pe^{rn} = 1{,}000e^{(0.08)(1)} = 1{,}000(1.0833) = \underline{\$1{,}083.29}$

b) $F = Pe^{rn} = 1{,}000e^{(0.08)(3)} = 1{,}000(1.2712) = \underline{\$1{,}271.25}$

7-19 **

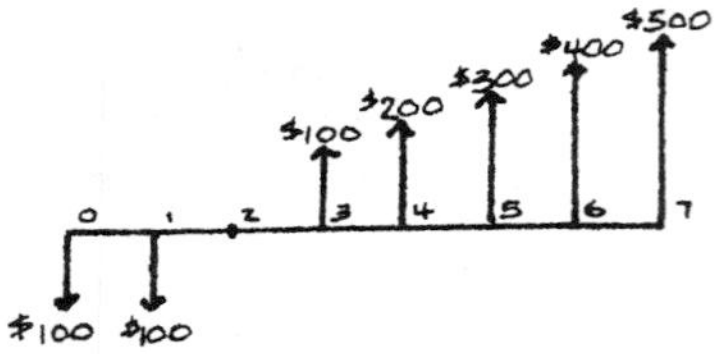

Using the tables for Uniform Gradients, solve for the Future Value at the end of year 7.

****

PV = 100(P/G,10%,7) − 100(P/A,10%,7) − 100

= 100(12.763) − 100(4.868) − 100 = 689.5

FV = 689.5(F/P,10%,7) = 689.5(1.949) = $1,343.84

7-20 **

If Jane invests $125,000 in a fund which will pay 12%, compounded quarterly, how much will she have in the fund at the end of 10 years? What effective interest rate is she earning on her investment?

****

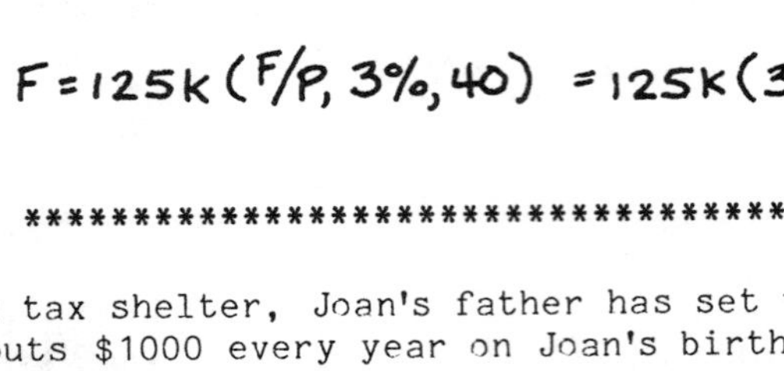

$i = \frac{12\%}{4} = 3\%$

$n = 10 \times 4 = 40$

F = 125K(F/P, 3%, 40) = 125K(3.262) = $407,750.

7-21 **

As a tax shelter, Joan's father has set up a trust fund into which he puts $1000 every year on Joan's birthday beginning on her first birthday. How much will be in the fund on Joan's 21st birthday if the account pays 6% compounded quarterly?

****

i = 6%/4 = 1½% per quarter

A = 1,000(A/F,1½%,4) = 1,000(.2444) = $244.40

F = 244.40(F/A,1½%,84) = 244.40(166.173) = $40,613.

PAYBACK PERIOD

** **7-22**

What is the major advantage of using Payback Period to compare alternatives?

****

Payback Period is easy and rapid. Also, it provides information on investment recovery time which may be important for companies with cash flow problems.

** **7-23**

For calculating payback period, when is the following formula valid?

$$\text{Payback Period} = \frac{\text{First Cost}}{\text{Annual Benefits}}$$

****

Valid when: a) There is a single first cost at time zero.
b) Annual Benefits = <u>Net</u> annual benefits after subtracting any annual costs
c) Net Annual Benefits are <u>uniform</u>

** **7-24**

Is the following statement True or False?

If two investors are considering the same project, the Payback Period will be longer for the investor with the higher Minimum Attractive Rate of Return (MARR).

****

Since Payback Period is generally the time to recover the investment, and ignores the MARR, it will be the same for both investors. The statement is False.

7-25 **

What is the Payback Period for a project with the following characteristics, if the Minimum Attractive Rate of Return (MARR) is 10%?

Useful Life	10 years
First Cost	$20,000
Salvage	2,000
Benefits	8,000/year
Maintenance	2,000 in year 1, then increasing by $500/year

****

Payback occurs when the sum of net annual benefits is equal to the first cost. Time value of money is ignored.

Year	Benefits		Costs		Net Benefits	Total Net Benefits
1	8,000	-	2,000	=	6,000	6,000
2	8,000	-	2,500	=	5,500	11,500
3	8,000	-	3,000	=	5,000	16,500
4	8,000	-	3,500	=	4,500	21,000 > 20,000

Payback Period = 4 years (Actually a little less)

7-26 **

Two mutually exclusive alternatives are found to be acceptable but A lasts twice as long as B. With no additional information given, which alternative is likely to have the shorter payback period? Why?

****

Alternative B is likely to have the shorter payback period. Since both are acceptable, they must return benefits greater than costs within the useful life. But B has only half as long a life, so it will almost certainly return benefits greater than cost before A and will therefore have a shorter payback period.

** **7-27**

A cannery is considering different modifications to some of their can fillers in two plants which have substantially different types of equipment. These modifications will allow better control and efficiency of the lines. The required investments amount to $135,000 in Plant A and $212,000 for Plant B. The expected benefits (which depend on the number and types of cans to be filled each year) are as follows:

Year	Plant A Benefits	Plant B Benefits
1	$ 73,000	$ 52,000
2	73,000	85,000
3	80,000	135,000
4	80,000	135,000
5	80,000	135,000

(a) Assuming MARR = 10%, which alternative is more profitable?

(b) Which alternative has the shortest payback period?

****

a) May be solved in various ways. Use PW method

$NPW_A = -135K + 73K(P/A, 10\%, 2) + 80K(P/A, 10\%, 3)(P/F, 10\%, 2)$
$= -135K + 73K(1.736) + 80K(2.487)(.8264) = 156,148.5$

$NPW_B = -212K + 52K(P/F, 10\%, 1) + 85K(P/F, 10\%, 2) + 135K(P/A, 10\%, 3)(P/F, 10\%, 2)$
$= -212K + 52K(.9091) + 85K(.8264) + 135K(2.487)(.8264)$
$= 182,976.8$

Therefore, modifications to plant B are more profitable

b)

	PLANT A		PLANT B	
YEAR	BENEFITS	CUMMULATIVE BENEFITS	BENEFITS	CUMMULATIVE BENEFITS
1	73,000	73,000	52,000	52,000
2	73,000	146,000*	85,000	137,000
3	80,000	226,000	135,000	272,000**

* The PBP of A is less than 2 years (1.85 years)

** The PBP of B is 2.55 years

Therefore, although not the most profitable, alternative A has the shortest payback period.

7-28 **

In this problem the minimum attractive rate of return is 10%. Three proposals are being considered.

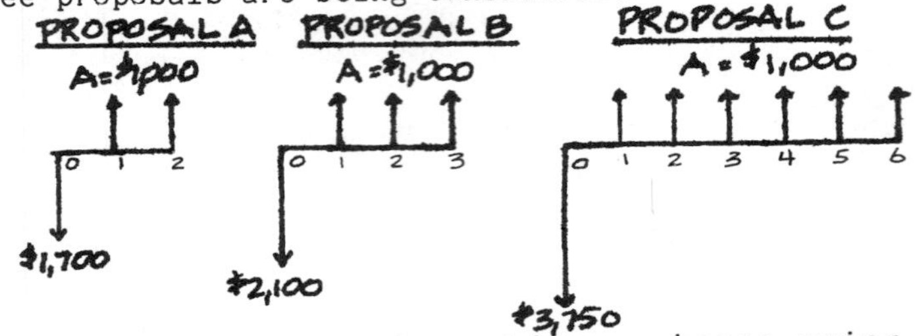

(a) Which proposal would you choose using Future Value analysis?

(b) How many years for Payback for each alternative? Which alternative would you choose?

****

PROPOSAL A EUAB = 1,000 − 1,700 (A/P, 10%, 2)
= 1,000 − 1,700 (.5762) = $20.5

PROPOSAL B EUAB = 1,000 − 2,100 (A/P, 10%, 3)
= 1,000 − 2,100 (.4021) = $155.6

PROPOSAL C EUAB = 1,000 − 3,750 (A/P, 10%, 6)
= 1,000 − 3,750 (.2296) = $139.0

a)
Proposal A : FW = 20.5 (F/A, 10%, 6) = 158.2
Proposal B : FW = 155.6 (F/A, 10%, 6) = 1,200.6
Proposal C : FW = 139 (F/A, 10%, 6) = 1,072.5
choose highest FW, Proposal B

b)
Proposal A : 1,700/1,000 = 1.7 years
Proposal B : 2,100/1,000 = 2.1 years
Proposal C : 3,750/1,000 = 3.75 years
choose quickest payback, proposal A

** **7-29**

Determine the Payback Period (to the nearest year) for the following project.

First Cost	$10,000
Maintenance	500 in Year 1, increasing by $200/year
Income	3,000/year
Salvage	4,000
Useful Life	10 years
MARR	10%

****

Year	Net Income	Sum
1	2,500	2,500
2	2,300	4,800
3	2,100	6,900
4	1,900	8,800
5	1,700	10,500 > 10,000

payback period = 5 years

** **7-30**

Determine the Payback Period (to the nearest year) for the following project:

Investment cost	$22,000
Maintenance cost	1,000/year
Uniform benefits	6,000/year
Overhaul costs	7,000 every 4 years
Salvage	2,500
Useful life	12 years
MARR	10%

****

Year	Σ Costs	Σ Benefits
0	22,000	–
1	23,000	6,000
2	24,000	12,000
3	25,000	18,000
4	33,000	24,000
5	34,000	30,000
6	35,000	36,000 ← Payback

Payback period = 6 years

BREAKEVEN

7-31 **

A machine that produces a certain piece must be turned off by the operator after each piece is completed. The machine "coasts" for 15 seconds after it is turned off, thus preventing the operator from removing the piece quickly before producing the next piece. An engineer has suggested installing a brake that would reduce the coasting time to 3 seconds.

The machine produces 50,000 pieces a year. The time to produce one piece is 1 minute 45 seconds, excluding coasting time. The operator earns $8.00 an hour and other direct costs for operating the machine are $4.00 an hour. The brake will require servicing every 500 hours of operation. It will take the operator 30 minutes to perform the necessary maintenance and will require $44.00 in parts and material. The brake is expected to last 7500 hours of operation (with proper maintenance) and will have no salvage value.

How much could be spent for the brake if the Minimum Attractive Rate of Return is 10% compounded annually?

****

Annual cost w/o brake : $50{,}000(2/60)(12) = \$20{,}000$

Annual cost w/ brake: $50{,}000(1.8/60)(12) = \$18{,}000$

Maintenance : $[(50{,}000(1.8/60))/500](.5(12)+44) = 150$

brake will last : $7{,}500/(5{,}000(1.8)/60) = 5$ yrs.

Max. amount : $(20{,}000 - 18{,}150)(P/A, 10\%, 5) = \$7{,}013.35$

7-32 **

A road can be paved with either asphalt or concrete. Concrete costs $15,000/km and lasts 20 years. What is the maximum which should be spent on asphalt if it only lasts 10 years? Annual maintenance costs for both pavements are $500/km. MARR = 12%.

****

Since maintenance is the same for both, it doesn't affect the answer. However, there is nothing wrong with including it.

$(15{,}000)(A/P, 12\%, 20) = P(A/P, 12\%, 10)$

$(15{,}000)(.1339) = P(.1770)$

$P = \dfrac{15{,}000(.1339)}{.1770} = \underline{\$11{,}347}$

** **7-33**

A proposed building may be roofed in either galvanized steel sheet or composition roofing. The composition roof costs $20,000 and must be replaced every 5 years at the same cost. The steel roof costs $28,000 but the useful life is unknown. Neither roof has any salvage value and no maintenance is needed. If the Minimum Attractive Rate of Return (MARR) equals 15%, what is the minimum life that the steel roof must have to make it the better alternative? (Report to the nearest whole year; don't bother interpolating.)

****

$EUAC_C = 20{,}000(A/P, 15\%, 5) = 20{,}000(.2983) = 5{,}966$

$EUAC_S = 28{,}000(A/P, 15\%, n)$

$(A/P, 15\%, n) = \frac{5966}{28{,}000} = 0.2131$

$(A/P, 15\%, 8) = .2229$

$(A/P, 15\%, 9) = .2096 \quad \therefore \underline{n = 9}$

** **7-34**

What is the breakeven capital cost for Project B compared to Project A if interest equals 10%?

Year	A	B
0	-$1000	?
1-5	+300/year	+200/year

****

NPW of A $= -1{,}000 + 300(P/A, 10\%, 5) = -1{,}000 + 300(3.791)$

$= 137.3$

NPW of B = NPW of A

$137.3 = P_B + 200(3.791) \rightarrow P_B = -620.90$

** **7-35**

What is the smallest acceptable annual income from a project which has a $70,000 investment cost and a $70,000 salvage value if the life is 15 years and the Minimum Attractive Rate of Return (MARR) is 20%?

****

Income $= 70{,}000(A/P, 20\%, 15) - 70{,}000(A/F, 20\%, 15)$

$= 70{,}000\, i = 70{,}000(.2)$

$= \underline{\$14{,}000}$

7-36 ***

A car rental agency has a contract with a garage to have them do major repairs (specified in the contract) for $450/car every six months. The car rental agency estimates that for $150,000, amortized at 8% interest for 20 years, and a salvage value of $60,000, they could have their own facility. They estimate that they could take care of their own car repairs in this facility at a cost of $200/car every six months. Ignoring taxes and other economic factors, what is the minimum number of cars needed to make the change feasible?

****

Let N = number of autos needed

$450 \times N = (150{,}000 - 60{,}000)(A/P, 4\%, 40) + 60{,}000(.04) + 200N$

$450N = 90{,}000 \times .0505 + 2{,}400 + 200N$

$250N = 6{,}945$

$N = \underline{27.78 \text{ or } 28 \text{ autos needed}}$

7-37 ***

Assume you need to buy some new automobile tires and you are considering purchasing either the "Econo-Ride," which costs $33.95 per tire, or the "Road King," which costs $65.50. Both tires are alike except that the "Road King" is more durable and will last longer. Regardless of which tire is purchased, balancing and installation costs are $1.50 per tire. The salesman says the "Econo-Ride" will last 20,000 miles. Assume a Minimum Attractive Rate of Return (MARR) of 6% and that you drive 10,000 miles per year.

(a) How many miles would the "Road King" have to last to make you indifferent in your choice?

(b) The salesman says the "Road King" will be on sale next week. If he also says the tire will last 30,000 miles, what would the sale price have to be to make you indifferent in your choice?

****

a) $4(1.5 + 33.95)(A/P, 6\%, \frac{20{,}000}{10{,}000}) = 4(1.5 + 65.50)(A/P, 6\%, N)$

$(A/P, 6\%, N) = .28859$, so $N = 4$, or 40,000 miles

b) $141.8\,(A/P, 6\%, 2) = (P + 6)(A/P, 6\%, \frac{30{,}000}{10{,}000})$,

so $P = \$50.18$

*** **7-38**

A soft drink company has researched the possibility of marketing a new low-calorie beverage, in a study region. The expected profits depend largely on the sales volume, and there is some uncertainty as to the precision of the sales-forecast figures. The estimated investment is $173,000 while the anticipated profits are $49,500 per year for the next 6 years. If the company's MARR = 15%, is the decision to invest sensitive to the uncertainty of the sales forecast, if it is estimated that in the worst case the profits will be reduced to $40,000 per year? What is the minimum volume of sales for the project to breakeven, if there is a profit of $6.70 per unit volume?

****

a) For an annual profit of $49,500

$NPW = 49{,}500(P/A, 15\%, 6) - 173{,}000 = 49{,}500(3.784) - 173{,}000$

$= +14{,}308$ (attractive)

b) For an annual profit of $40,000

$NPW = 40{,}000(P/A, 15\%, 6) - 173{,}000 = 40{,}000(3.784) - 173{,}000$

$= -21{,}640$ (not attractive)

Therefore the decision is sensitive to the expected variations in sales or profits.

The breakeven: NPW = 0

$NPW = 0 = x(P/A, 15\%, 6) - 173{,}000$ where x = min $ profit

$$x = \frac{173{,}000}{(P/A, 15\%, 6)} = \frac{173{,}000}{3.784} = \$45{,}718.8$$

$$\text{in volume units} = \frac{\$45{,}718.8}{\$6.70/\text{unit}} = 6{,}824 \text{ volume units}$$

*** **7-39**

A machine, costing $2000 to buy and $300 per year to operate, will save labor expenses of $650 per year for 8 years. If the interest rate is 10%, what is the minimum salvage value (after 8 years) at which the machine is worth purchasing?

****

$NPW = -2{,}000 + 350(P/A, 10\%, 8) + S(P/F, 10\%, 8) = 0$

$= -2{,}000 + 350(5.335) + S(.4665) = 0$

$-132.75 + .4665S = 0$

$$S = \frac{132.75}{.4665} = \$284.57$$

7-40 **

The PARC Company can purchase gismoes to be used in building whatsits for $90 each. PARC can manufacture their own gismoes for $7000 per year overhead cost plus $25 direct cost for each gismoe, provided they purchase a gismoe maker for $100,000. PARC expects to make whatsits using gismoes for 10 years. The gismoe maker should have a salvage value of $20,000 after 10 years. PARC uses 12% as its minimum attractive risk rate. At what annual production rate should PARC make their own gismoes?

****

Equivalent Uniform Annual Cost Solution:

$EUAC_{Buy} = \$90N$ where N = annual quantity

$EUAC_{Make} = 100{,}000(A/P, 12\%, 10) + 7{,}000 + 25N - 20{,}000(A/F, 12\%, 10)$

$= 100{,}000(0.1770) + 7{,}000 + 25N - 20{,}000(0.0570)$

$= 23{,}560 + 25N$

For breakeven:

$EUAC_{Buy} = EUAC_{make}$

$90N = 23{,}560 + 25N \qquad N = \frac{23{,}560}{65} = 362.5$

This indicates they should be bought at 362/year or less and made at 363/year or more.

8
DEPRECIATION COMPUTATIONS

** **8-1**

Some seed cleaning equipment was purchased in 1978 for $8500 and is depreciated by the Double Declining Balance (DDB) method for an expected life of 12 years. What is the book value of the equipment in 1984? Original salvage value was estimated to be $2500 at the end of 12 years.

****

$$\text{Book Value} = P\left(1-\frac{2}{N}\right)^n$$

$$= 8{,}500\left(1-\frac{2}{12}\right)^6 = \$2{,}846.63$$

This can be checked by doing the year-by-year computations:

YEAR	DDB
1977	$\frac{2}{12}(8{,}500 - 0) = 1{,}416.67$
1978	$\frac{2}{12}(8{,}500-1{,}416.67) = 1{,}180.56$
1979	$\frac{2}{12}(8{,}500-2{,}597.23) = 983.80$
1980	$\frac{2}{12}(8{,}500-3{,}581.03) = 819.83$
1981	$\frac{2}{12}(8{,}500-4{,}400.86) = 683.19$
1982	$\frac{2}{12}(8{,}500-5{,}084.05) = \underline{569.32}$

Book Value $= 8{,}500 - 5{,}653.37 = \underline{\$2{,}846.63}$

8-2 **

It is expected that an asset will cost $1750 when purchased in 1986. It is further expected to have a salvage value of $250 at the end of its five year depreciable life. Calculate complete depreciation schedules giving the depreciation charge, D(n), and end-of-year book value, B(n), for straight line (SL), sum of the years digits (SYD), double declining balance (DDB), and accelerated cost recovery (ACRS) depreciation methods. Use ACRS percentages of 20%, 32%, 24%, 16%, and 8%, for years 1 through 5, respectively.

****

	SL		SYD		DDB		ACRS	
n	D(n)	B(n)	D(n)	B(n)	D(n)	B(n)	D(n)	B(n)
0		1,750		1,750		1,750		1,750
1	300	1,450	500	1,250	700	1,050	350	1,400
2	300	1,150	400	850	420	630	560	840
3	300	850	300	550	252	378	420	420
4	300	550	200	350	128	250	280	140
5	300	250	100	250	0	250	140	0

8-3 **

A pump costs $1000 and has a salvage value of $100 after a life of five years. Using the double rate declining balance depreciation method, determine:

(a) The depreciation in the first year.
(b) The tax write-off in the first year based on a 40% tax rate.
(c) The book value after five years.
(d) The book value after five years if the salvage was only $50.

****

a) Rate $= \frac{200\%}{5} = 40\% = .4$

$1{,}000(.4) = \underline{\$400}$.

b) Write-off for tax = 400

Tax reduction = 160

$400(.4) = \underline{\$160}$.

c) $B.V. = \max.\{S.V.;\ 1{,}000(.6)^5\} = \max\{100, 77.76\} = \underline{\$100}$.

d) $B.V. = \max.\{S.V.;\ 1{,}000(1-.4)^5\} = \max.\{50, 77.76\} = \underline{\$77.76}$.

** **8-4**

A new machine costs $12,000 and has a $1200 salvage value after using it for eight years. Prepare a year-by-year depreciation schedule by the Double Declining Balance (DDB) method with a switch to Straight Line if appropriate.

****

DDB Depr $= \frac{2}{N}(P - D_{acc})$

YEAR	1	2	3	4	5	6
DEPR.	3,000	2,250	1,688	1,266	949	712

YEAR	7	8	Total
DEPR.	534	401	$10,800

Total Depreciation = $10,800 when N = 8

** **8-5**

If you were to use Double Declining Balance depreciation for an asset which cost $100,000 and had an estimated salvage value of $5000 and an 8-year useful life, in which year would you switch to straight line depreciation?

****

YEAR	DDB DEPREC.	REMAINING BOOK VALUE	STRAIGHT LINE DEPREC. FOR NEXT YEAR
1	$\frac{2}{8}(100{,}000) = 25{,}000$	75,000	$\frac{75{,}000 - 5{,}000}{7}$ = $10,000
2	.25(75,000) = 18,750	56,250	51,250/6 = $ 8,542
3	.25(56,250) = 14,063	42,188	37,188/5 = $ 7,438
4	.25(42,188) = 10,547	31,641	26,641/4 = $ 6,660
5	.25(31,641) = 7,910	23,731	18,731/3 = $ 6,244
6	.25(23,731) = 5,933	17,798	

starting in year 6, convert to straight line depreciation. year 6 straight line depreciation would be $6,244 vs a DDB method $5933.

8-6 **

Suppose the tax laws are changed in 1989 as a result of the President's new economic policies. One class of property is assigned a 6-year depreciable life. The depreciation schedule is based on 150% declining balance switching to straight line, with a full year's depreciation in the first year.

Compute the percentage recovery factors to be applied to the purchase price for determining the depreciation charge in each year. Book value at the end of the depreciable life is to be zero. Round to the nearest percent.

****

YEAR	150% Declining Bal.	S.L.	% RECOVERY FACTOR
1	$\frac{1.5}{6}(100\%) = 25\%$	$\frac{100\%}{6} = 17\%$	25%
2	$\frac{1.5}{6}(100 - 25) = 19\%$	$\frac{75\%}{5} = 15\%$	19%
3	$\frac{1.5}{6}(75 - 19) = 14\%$ switch (either one)	$\frac{56\%}{4} = 14\%$	14%
4	$\frac{1.5}{6}(56 - 14) = 11\%$	14%	14%
5	$\frac{1.5}{6}(42 - 11) = 8\%$	14%	14%
6	$\frac{1.5}{6}(31 - 8) = 6\%$	14%	14%

Total depr. = 100%

Don't need to compute. Note that DB depreciates only 83% of original value.

8-7 **

A piece of machinery costs $5000 and has an anticipated $1000 resale value at the end of its five year useful life. Compute the depreciation schedule for the machinery by the Sum-Of-Years-Digits method.

****

Sum-of-years-digits $= \frac{n}{2}(n+1) = \frac{5}{2}(6) = 15$

1st year depreciation $= \frac{5}{15}(5{,}000 - 1{,}000) = \$1{,}333$

2nd year depreciation $= \frac{4}{15}(5{,}000 - 1{,}000) = \$1{,}067$

3rd year depreciation $= \frac{3}{15}(5{,}000 - 1{,}000) = \$800$

4th year depreciation $= \frac{2}{15}(5{,}000 - 1{,}000) = \$533$

5th year depreciation $= \frac{1}{15}(5{,}000 - 1{,}000) = \$267$

$4,000

** **8-8**

We are considering the purchase of second-hand minicomputer at a cost of $10,500, with an estimated salvage value of $500 and a projected useful life of four years. Interest is 10%.

Determine:
- (a) Sum Of Years Digits (SOYD) depreciation
- (b) Double Declining Balance (DDB) depreciation
- (c) Double Declining Balance with conversion to Straight Line depreciation
- (d) Using present worth analysis, determine the best alternative between DDB with conversion to SL, and SOYD depreciation.

****

P = 10,500, S = 500, n = 4, i = 10%

year	(a) SOYD	(b) DDB	(c) DDB/SL
1	4,000	5,250	5,250
2	3,000	2,625	2,625
3	2,000	1,312.50	1,312.50
4	1,000	656.25	812.50
	10,000	9,843.75	10,000

DDB/SL 3yr. $SL = \frac{[10{,}500 - (5{,}250 + 2{,}625) - 500]}{2} = 1{,}062.50$

4yr. $SL = \frac{[10{,}500 - (5{,}250 + 2{,}625 + 1{,}312.50)] - 500}{1} = 812.50$

d) SOYD(PW) = 4,000(P/A,10%,4) − 1,000(P/G,10%,4)
= 4,000 (3.170) − 1,000(4.378) = 8,302.

DDB/SL(PW) = 5,250(P/F,10%,1) + 2,625 (P/F,10%,2)
+ 1,312.50(P/F,10%,3) + 812.50(P/F,10%,4)
= 5,250(.9091) + 2,625(.8264) + 1,312.50(.7513)
+ 812.50 (.6830) = 8,483.09

Comparing SOYD & DDB/SL, <u>DDB/SL is best</u>

8-9

A machine costs $5000 and has an estimated salvage value of $1000 at the end of 5 years useful life. Compute the depreciation schedule for the machine by

(a) Straight Line (SL)
(b) Double Declining Balance (DDB)
(c) Sum Of Years Digits (SOYD)

a) $SL = \frac{P-S}{N} = \frac{5{,}000 - 1{,}000}{5} = \$800$

$\Sigma D_c = 800 \times 5 = 4{,}000$

$S = 5{,}000 - 4{,}000 = \$1{,}000$

b) $DDS = \frac{2}{N}[P - \Sigma D_c(t)]$

D_c (1st yr.) $= \frac{2}{5}[5{,}000 - 0] =$	\$2,000	→ \$2,000
D_c (2nd yr.) $= \frac{2}{5}[5{,}000 - 2{,}000] =$	\$1,200	→ \$1,200
D_c (3rd yr.) $= \frac{2}{5}[5{,}000 - 3{,}200] =$	\$720	→ \$720
D_c (4th yr.) $= \frac{2}{5}[5{,}000 - 3{,}920] =$	\$432	→ \$80
D_c (5th yr.) $= \frac{2}{5}[5{,}000 - 4{,}352] =$	\$259.2	→ 0
	4,611.2	4,000
	− 4,000.	
	611.2	

c) $SOYD = \frac{N}{2}(N+1) = \frac{5}{2}(6) = 15$

$D_c(1st) = \frac{5}{15}(5{,}000 - 1{,}000) = \$1{,}333$

$D_c(2nd) = \frac{4}{15}(4{,}000) = \$1{,}067$

$D_c(3rd) = \frac{3}{15}(4{,}000) = \$800$

$D_c(4th) = \frac{2}{15}(4{,}000) = \$533$

$D_c(5th) = \frac{1}{15}(4{,}000) = \$267$

4,000

** **8-10**

An asset has a purchase cost of $100,000 and a depreciable life of 10 years.

Part One: Calculate the year 2 depreciation charges for each of depreciation methods listed below, assuming a zero salvage value.

(a) Straight Line (b) Double Declining Balance
(c) Sum Of Years Digits (d) 150% Declining Balance

Part Two: Calculate the year 2 depreciation charges for each of the depreciation methods listed in Part One, assuming an end of depreciable life salvage value of $80,000.

****

Part One:

a) $SL = \frac{1}{N}(P-0) = \frac{100{,}000}{10} = \underline{10{,}000} \leftarrow$

b) $DDB_1 = \frac{2}{10}(100{,}000) = 20{,}000$

$DDB_2 = \frac{2}{10}(100{,}000 - 20{,}000) = \underline{16{,}000} \leftarrow$

c) $SOYD = \frac{10(10+1)}{2} = 55\ ;\ \frac{10+9}{55}(100{,}000) = \underline{16{,}363.64} \leftarrow$

d) $150\%\ DB_1 = \frac{1.5}{10}(100{,}000) = 15{,}000$

$150\%\ DB_2 = \frac{1.5}{10}(100{,}000 - 15{,}000) = \underline{12{,}750} \leftarrow$

Part Two:

a) $SL = \frac{1}{10}(100{,}000 - 80{,}000) = \underline{2{,}000} \leftarrow$

b) $DDB_1 = \frac{2}{10}(100{,}000) = 20{,}000$

book value at year two = 80,000 ∴ totally depreciated

$DDB_2 = \underline{0} \leftarrow$

c) $SOYD = \frac{10+9}{55}(100{,}000 - 80{,}000) = \underline{3{,}272.73} \leftarrow$

d) $150\%\ DB_1 = \frac{1.5}{10}(100{,}000) = 15{,}000$

$150\%\ DB_2 = \frac{1.5}{10}(100{,}000 - 15{,}000) = 12{,}750$ but book value at year 2 would be 72,250

$\therefore 150\%\ DB_2 = 85{,}000 - 80{,}000 = \underline{5{,}000} \leftarrow$

8-11 **

A used piece of depreciable property was bought for $20,000. If it has a useful life of 10 years and a salvage value of $5000, how much will it be depreciated in the 9th year, using the 150% declining balance schedule?

****

$$\text{Depr.} = \frac{1.5P}{N}\left(1-\frac{1.5}{N}\right)^{n-1} = \frac{1.5(20{,}000)}{10}\left(1-\frac{1.5}{10}\right)^{9-1} = \$817.50$$

check BV at end of 8th year

$$BV = P\left(1-\frac{1.5}{N}\right)^{n} = 20{,}000\left(1-\frac{1.5}{10}\right)^{8} = 5{,}449.80$$

But salvage value is \$5,000 ∴ in 9th year you can only depreciate 5,449.80 − 5,000 = <u>\$449.80</u>. The \$817.50 would have brought it below salvage value of \$5,000.

8-12 **

A front-end loader cost $70,000 and has an estimated salvage value of $10,000 at the end of 5 years useful life. Compute the depreciation schedule, and book value, to the end of the useful life of the tractor by

(a) Straight Line (SL)
(b) Sum Of Years Digits (SOYD)
(c) Double Declining Balance (DDB) with conversion to straight line if necessary.

****

a) straight line (SL)

$$D_c = (P-S)/N = (70{,}000-10{,}000)/5 = 12{,}000 \text{ per yr.}$$

YR	Dc	BV
1	12,000	70,000 − 12,000 = 58,000
2	12,000	70,000 − 24,000 = 46,000
3	12,000	70,000 − 36,000 = 34,000
4	12,000	70,000 − 48,000 = 22,000
5	12,000	70,000 − 60,000 = 10,000
	Σ = 60,000	

b) sum of years digit (SOYD)

$SOYD = \frac{N}{2}(N+1) = \frac{5}{2}(5+1) = 15$

$D_c = \frac{N}{SOYD}(P-S)$

YR	D_c		BV
1	5/15 (70,000 − 10,000)	= 20,000	70,000 − 20,000 = 50,000
2	4/15 (60,000)	= 16,000	70,000 − 36,000 = 34,000
3	3/15 (60,000)	= 12,000	70,000 − 48,000 = 22,000
4	2/15 (60,000)	= 8,000	70,000 − 56,000 = 14,000
5	1/15 (60,000)	= 4,000	70,000 − 60,000 = 10,000
		Σ = 60,000	

c) double declining balance (DDB)

$DDB = \frac{2}{N}[P - \Sigma D_c\{\text{to date}\}]$

YR	DDB			BV
1	2/5 (70,000 − 0) =	28,000		70K − 28,000 = 42,000
2	2/5 (70,000 − 28,000) =	16,800	44,800	70K − 44,800 = 25,200
3	2/5 (70,000 − 44,800) =	10,080	54,880	70K − 54,880 = 15,120
4	2/5 (70,000 − 54,880) =	6,048	60,928	70K − 60,928 = 9,072 ←
5	2/5 (70,000 − 60,928) =	3,628.8	64,556.8	70K − 64,556.8 = 5,443.2

depreciated to end of useful life means to convert to straight line:

3YR: $D_c(SL) = [BV(BOY) - S]/RUL = [25,000 - 10,000]/3 = 5,066.67$
< 10,800

4YR: $D_c(SL) = [BV(BOY) - S]/RUL = [15,120 - 10,000]/2 = 2,560/\text{yr}$

YR	DDB	BV = P − ΣDc
1	28,000	70,000 − 28,000 = 42,000
2	16,800	70,000 − 44,800 = 25,200
3	10,080	70,000 − 54,880 = 15,120
4	2,560	70,000 − 57,440 = 12,560
5	2,560	70,000 − 60,000 = 10,000
	60,000	

8-13 **

To meet increased sales, 10 new special delivery trucks will be purchased by a large dairy. Each one costs $18,000. Compute the depreciation schedule for each truck, by the following methods:

(a) Straight line, declining balance and sum-of-years digits; assuming a salvage value of $2000 for each truck at the end of 4 years.

(b) By the Accelerated Cost Recovery System (ACRS) method, if this property class is depreciable in 3 years, and the following depreciation rates should be applied for years 1, 2, and 3, respectively: 33%, 45%, and 22%.

****

a) - Straight line

Depr. Charge $= \frac{1}{N}(P-S) = \frac{1}{4}(18{,}000-2{,}000) = 4{,}000$ \$/year

- Double declining balance

Depr. charge $= \frac{2}{N}(\text{Book Value}) = \frac{2}{4}(\text{Book Value}) = \frac{1}{2}(\text{Book Value})$

Year	B.V. Before Depreciation	Depreciation Charge	B.V. After Depreciation
1	18,000	9,000	9,000
2	9,000	4,500	4,500
3	4,500	2,250	2,250
4	2,250	1,125	1,125 *

* Note this is lower than the S value. In practice this would not be permissible.

- Sum-of-Years digits

Depreciation charges $= \frac{\text{Remaining useful life at beginning of year}}{\text{SYOD} \rightarrow \frac{N}{2}(N+1)}(P-S)$

$\frac{N}{2}(N+1) = \frac{4}{2}(4+1) = 10$

Year	SOYD Depreciation
1	$\frac{4}{10}(18{,}000-2{,}000) = 6{,}400$
2	$\frac{3}{10}(16{,}000) = 4{,}800$
3	$\frac{2}{10}(16{,}000) = 3{,}200$
4	$\frac{1}{10}(16{,}000) = 1{,}600$

b) by ACRS

Year	ACRS Depreciation
1	0.33 (18,000) = 5,940
2	0.45 (18,000) = 8,100
3	0.22 (18,000) = 3,960
4	0

** **8-14**

A lumber company purchased a tract of timber for $70,000. The value of the 25,000 trees on the tract was estimated to be $50,000. The value of the land was estimated to be $20,000. In the first year of operation, the lumber company cut down 5000 trees. What was the depletion allowance for the year?

****

For standing timber only cost depletion (not percentage depletion) is permissible. 5,000/25,000 = 0.20. Thus 1/5 of the tract was depleted. Land is not considered depletable, only the timber, which is valued at a total of $50,000. Therefore, the first year's depletion allowance would be

0.20($50,000) = $10,000.

** **8-15**

A stamping machine cost $9000, has a useful life of 6 years, and an estimated salvage value of $1500. Using double declining balance depreciation, compute both the depreciation schedule and the book value for each year. Make the conversion to straight line depreciation if this is advantageous.

****

$$\text{Dep.} = \frac{2}{6}(\text{B.V.})$$

YEAR	DEPRECIATION	BOOK VALUE
0	–	9,000
1	3,000	6,000
2	2,000	4,000
3	1,333	2,667
4	889	1,778
5	~~593~~ 278	~~1,185~~ 1,500
6	0	1,500

← Book value can not drop below salvage value. Depreciation stops.

8-16 **

A machine was purchased two years ago for $50,000 and had a depreciable life of five years. The owner is considering an offer to sell the machine for $25,000. For each of the depreciation methods listed, fill in the table below to determine the depreciation for year 2, and the book value at the end of year 2.

	Depreciation For Year 2	End of Year 2 Book Value
Sum-Of-Years Digits (SOYD)		
Stright Line (SL)		
Double Declining Balance (DDB)		
Accelerated Cost Recovery System (ACRS)		

****

	Depreciation For Year 2	End of Year 2 Book Value
Sum-Of-Years Digits (SOYD)	13,333.$\overline{33}$	20,000
Stright Line (SL)	10,000	30,000
Double Declining Balance (DDB)	12,000	18,000
Accelerated Cost Recovery System (ACRS)	16,000	24,000

Straight Line:

depreciation - $\frac{1}{N}(P-S) = \frac{1}{5}(50{,}000-0) = 10{,}000$ per year

cumulative depr. = 20,000

book value = P - cumm. depr = 50,000 - 20,000

SOYD:

$SOYD = \frac{5 \cdot 6}{2} = 15$

depreciation yr. 1 $= \frac{5}{15}(50{,}000) = 16{,}666.\overline{66}$

depreciation yr. 2 $= \frac{4}{15}(50{,}000) = 13{,}333.\overline{33}$

cumulative depr. = 30,000
book value = 50,000 - 30,000 = 20,000

DDB:

depreciation year 1 = $\frac{2}{N}$ (Book Value) = $\frac{2}{5}$(50,000) = 20,000
depreciation year 2 = $\frac{2}{5}$(50,000 - 2000) = 12,000
cumulative depr. = 32,000
book value = 50,000 - 32,000 = 18,000

ACRS:

depreciation year 1 = (20%)(50,000) = 10,000
depreciation year 2 = (32%)(50,000) = 16,000
cumulative depreciation = 26,000
book value = 50,000 - 26,000 = 24,000

** **8-17**

In the production of beer, a final filtration is given by the use of "Kieselguhr" or diatomaceous earth, which is composed of the fossil remains of minute aquatic algae, a few microns in diameter and composed of pure silica. A company has purchased a property for $840,000 which contains an estimated 60,000 tons. Compute the depreciation charges for the first three years, if a production (or extraction) of 3000 tons, 5000 tons, and 6000 tons are planned for years 1, 2, and 3, respectively. Use the cost-depletion method, assuming no salvage value for the property.

****

Total diatomaceous earth in property = 60,000 tons
Cost of property = $840,000
Then, depletion allowance per ton extracted = $\frac{\$840{,}000}{60{,}000 \text{ tons}}$ = 14 $/ton

year	tons of diatomaceous earth extracted	depreciation charge
1	3,000	3,000 x 14 = $42,000
2	4,000	4,000 x 14 = $56,000
3	5,000	5,000 x 14 = $70,000

8-18 **

On January 1 you purchase a production press for $10,500. The press has a 5-year useful life and a $500 expected salvage value.

PART 1. Compute the yearly depreciation using:
- (a) Straight line depreciation
- (b) Sum of years digits depreciation
- (c) Double declining balance depreciation
- (d) Double declining balance depreciation with conversion to straight line
- (e) Accelerated Cost Recovery System depreciation, based on the 5-year property class and ACRS percentages of 20%, 32%, 24%, 16%, and 8%, respectively

PART 2. If the estimated salvage value was $1000, rather than $500, what would the DDB depreciation be in year 5?

PART 3. The ACRS depreciation percentages are based on DDB in the early years. Why then is the ACRS method first year's depreciation so much lower than when using DDB?

****

PART 1:

a) Straight line -

annual depreciation charge $= \frac{\$10,500 - \$500}{5} = \$2,000$

b) Sum of years -

sum of years digits $= \frac{5}{2}(5+1) = 15$

1st year $= \frac{5}{15}(10,500 - 500) = \$3,333$

2nd year $= \frac{4}{15}(10,000) = \$2,667$

3rd year $= \frac{3}{15}(10,000) = \$2,000$

4th year $= \frac{2}{15}(10,000) = \$1,333$

5th year $= \frac{1}{15}(10,000) = \$667$

check $\$10,000$

c) Double Declining Balance -

1st year $= \frac{2}{5}(10,500) = \$4,200$

2nd year $= \frac{2}{5}(10,500 - 4,200) = \$2,520$

3rd year $= \frac{2}{5}(6,300 - 2,520) = \$1,512$

4th year $= \frac{2}{5}(3,780 - 1,512) = \$907$

5th year $= \frac{2}{5}(2,268 - 907) = \$544$

d) Double Declining Balance with Conversion to Straight Line -

In method (c) the remaining book value at end

year 5 would be $817. This is greater than the estimated $500 salvage value and clearly conversion to straight line is warrented in the year when straight line depreciation of the remaining "book value" would be greater than the DDB depreciation.

YEAR	DDB DEPREC	BOOK VALUE	STRAIGHT LINE DEPREC. FOR NEXT YR
1	$4,200	$6,300	(6,300 − 500)/4 = $1,450
2	2,520	3,780	(3,780 − 500)/3 = $1,093
3	1,512	2,268	(2,268 − 500)/2 = $884
4	907	1,361	(1,361 − 500)/1 = $861
5	544		

Conversion to straight line depreciation in year 5 would provide an $861 depreciation allowance which is greater than the $544 sum of years digits depreciation allowance. For years 1-4 it is more advantageous to continue using DDB.

e) ACRS -

year 1 = .20(10,500) = $2,100
year 2 = .32(10,500) = $3,360
year 3 = .24(10,500) = $2,520
year 4 = .16(10,500) = $1,680
year 5 = .08(10,500) = $840

YEAR	SL	SOYD	DDB	DDB w/SL	ACRS
1	2,000	3,333	4,200	4,200	2,100
2	2,000	2,667	2,520	2,520	3,360
3	2,000	2,000	1,512	1,512	2,520
4	2,000	1,333	907	907	1,680
5	2,000	667	544	861	840

PART 2:

The book value at the end of year 4 would still

be $1,361. Therefore in year 5 $361 depreciation allowance would be allowed.

PART 3:

The ACRS depreciation schedule assumes the property was acquired in the middle of the first year.

8-19 **

Given the following two depreciation schedules, determine which one is best. Use an interest rate of 6% per annum.

Year	SOYD	DDB/SL
1	$16,000	$19,000
2	14,000	14,250
3	12,000	10,688
4	10,000	8,015
5	8,000	6,012
6	6,000	4,678
7	4,000	4,678
8	2,000	4,678
	$72,000	$72,000

****

The "best" depreciation schedule would be the one that depreciates the asset most rapidly. More precisely, it is the depreciation schedule whose Present Worth is greatest.

SOYD:

$P = 16{,}000(P/A, 6\%, 8) - 2{,}000(P/G, 6\%, 8)$

$= 16{,}000(621) - 2{,}000(19.842) = \$59{,}676$

DDB/SL:

$P = 19{,}000(P/F, 6\%, 1) + 14{,}250(P/F, 6\%, 2) + 10{,}688(P/F, 6\%, 3)$

$+ 8{,}015(P/F, 6\%, 4) + 6{,}012(P/F, 6\%, 5) + 4{,}678(P/F, 6\%, 3)(P/F, 6\%, 5)$

$= 19{,}000(.9434) + 14{,}250(.8900) + 10{,}688(.8396) + 8{,}015(.7921)$

$+ 6{,}012(.7473) + 4{,}678(2.673)(.7473) = \$59{,}766.65$

∴ choose Double Declining Balance with conversion to straight line

** **8-20**

Given two depreciation schedules for a small microprocessor, which method of depreciation should be selected, based upon i = 12%?

Year	DDB	SOYD
1	$333.33	$285.71
2	222.22	238.09
3	148.15	190.48
4	98.77	142.86
5	98.77	95.24
6	98.76	47.62

****

A firm wants to depreciate its assets as rapidly as possible. In situations where the choice between methods is not obvious, then computations are required. Select the depreciation method with the largest Present Worth of depreciation charges.

DDB w/conv. SL

YR	DDB		P/F, 12%, n		
1	333.33	×	.8929	=	297.63
2	222.22	×	.7972	=	177.15
3	148.15	×	.7118	=	105.45
4	98.77	×	.6355	=	62.77
5	98.77	×	.5674	=	56.04
6	98.76	×	.5066	=	50.03
	1,000.00				749.07

SOYD

YR	SOYD		P/F, 12%, n		
1	285.71	×	.8929	=	255.11
2	238.09	×	.7972	=	189.80
3	190.48	×	.7118	=	135.58
4	142.86	×	.6355	=	90.79
5	95.24	×	.5674	=	54.04
6	47.62	×	.5066	=	24.12
	1,000.00				749.44

∴ Select SOYD

8-21

Adventure Airlines recently purchased a new baggage crusher for \$50,000. It is expected to last for 14 years and have an estimated salvage value of \$8000. Determine the depreciation charge on the crusher for the third year of its life and the book value at the end of 8 years, using sum-of-digits depreciation.

(a) SOYD depreciation for 3rd year.

Sum of Years digits $= \frac{n}{2}(n+1) = \frac{14}{2}(14+1) = 105$

3rd year depreciation $= \dfrac{\text{remaining life beginning of year}}{\Sigma \text{ years digits}}(P-S)$

$= \frac{12}{105}(50{,}000-8{,}000) = \underline{4{,}800}$

(b) Book Value at end of 8 years

Σ 8 years of depreciation $= \frac{14+13+12+11+10+9+8+7}{105}(50{,}000-8{,}0$

$= \frac{84}{105}(42{,}000) = 33{,}600$

Book Value = Cost − Depreciation to date $= 50{,}000 - 33{,}600$

$= \underline{16{,}400}$

9
INCOME TAXES

** **9-1**

Municipal (government) bonds are tax-exempt securities, that is, a taxpayer who owns a municipal bond pays no federal taxes on the bond. As a result, investing in municipal bonds is often attractive to taxpayers who have high federal tax rates.

Assume a taxpayer is considering the purchase of either a municipal bond or a corporate bond (which is subject to taxation), both of which are summarized below. Assume his Minimum Attractive Rate of Return is 8% after taxes and that the taxpayer's (federal) capital gain tax rate is 40% of his (federal) income tax rate, that is, $d_G = 0.4d$. What would the taxpayer's (federal) income tax rate, d, have to be for him to be indifferent in choosing between the two bonds? Assume interest payments are made annually. Ignore state taxes.

	Municipal Bond	Corporate Bond
Face Value (Denomination)	$5000	$5000
Market Price	3770	2010
Annual Interest Rate	6%	10%
Maturity	10 yrs	10 yrs

****

Any measure of worth could be used. The solution here sets the net present values equal to each other then solve for d.

$-3{,}770 + 0.06(5{,}000)(P/A, 8\%, 10) + 5{,}000(P/F, 8\%, 10)$

$= -2{,}010 + 0.1(5{,}000)(1-d)(P/A, 8\%, 10) + 2{,}010(P/F, 8\%, 10)$

$+ (5{,}000 - 2{,}010)(1 - .4d)(P/F, 8\%, 10)$

$558.99 = 3{,}661.01 - 3{,}909.02\, d$

$d = 0.794$

9-2 ***

A tool costing $300 has no salvage value. Its resultant cash flow before tax is shown below.

Year	Cash Flow Before Tax	SOYD Write-Off	Effect On Taxable Income	Effect On Cash Flow For Tax	Cash Flow After Tax
0	-$300				
1	+100				
2	+150				
3	+200				

The tool must be depreciated over 3 years according to the Sum Of Years' Digits method. The tax rate is 50%.

(a) Fill in the four columns in the table.

(b) What is the internal rate of return after tax?

****

a)

Year	Cash Flow Before Tax	SOYD Write-Off	Effect On Taxable Income	Effect On Cash Flow For Tax	Cash Flow After Tax
0	-300				-300
1	+100	-150	-50	+25	+125
2	+150	-100	+50	-25	+125
3	+200	-50	+150	-75	+125

b) $NPW_{AT} = -300 + 125(P/A, i\%, 3) = 0$

$(P/A, i\%, 3) = \frac{300}{125} = 2.4$

$IRR = i \approx 12\%$ $(\approx 12.05\%)$

9-3 ***

Pete put $10,000 in a fund that paid him $250 every three months. Because of changes in the tax laws to close certain loopholes, the fund was liquidated at the end of 3 years and Pete received $12,000 as his share, in addition to the normal $250 payment. Treating the $250 as taxable income and the $2000 as a capital gain (taxable income = 40% of capital gain), what was Pete's nominal annual after-tax rate of return (that is, with quarterly compounding) if his incremental income tax rate is 38%? Assume the taxes are paid when the money is received. Hint: To find an **i** for the first try, use $(P/F,i,n) \approx P/(F + nA)$.

****

Quarter	BTCF	Tax. Inc.	Taxes @ 38%	ATCF
0	-10,000	–	–	-10,000
1-12	250	250	95	155
S	12,000	800	304	11,696

$NPW = 155(P/A, i, 12) + 11{,}696(P/F, i, 12) - 10{,}000 = 0$

Trial i: $(P/F, i, 12) \approx \dfrac{10{,}000}{11{,}696 + 12(155)} = 0.738$ From tables $i \approx 2\tfrac{1}{2}\%$

Try $i = 2\tfrac{1}{2}\%$: $155(10.258) + 11{,}696(0.7436) - 10{,}000 = 287$

Try $i = 3\%$: $155(9.954) + 11{,}696(0.7014) - 10{,}000 = -254$

$i = 2\tfrac{1}{2} + \tfrac{1}{2}\left(\dfrac{287}{287+254}\right) = 2.77\%$

Nominal Annual Rate $= 4 \times 2.77 = \underline{11.1\%}$

** **9-4**

A company, whose earnings put them in the 46% tax schedule, is considering purchasing a piece of equipment for $25,000. The equipment has a straight line depreciation, a useful life of 4 years and a salvage value of $5000. It is estimated that the equipment will increase the company's earnings by $8000 for each of the 4 years. Should the equipment be purchased? Use an interest rate of 10%.

****

Year	BTCF	Deprec.	ΔTI	46% ΔIT	ATCF
0	-25,000	0			-25,000
1	+ 8,000	5,000	+ 3,000	1,380	6,620
2	+ 8,000	5,000	+ 3,000	1,380	6,620
3	+ 8,000	5,000	+ 3,000	1,380	6,620
4	+ 8,000	5,000	+ 3,000	1,380	6,620
	+ 5,000	–			5,000

S/L Depr. $= \tfrac{1}{4}(25{,}000 - 5{,}000) = \$5{,}000$

Find net present worth : if > 0 then good deal

$6{,}620(P/A, 10\%, 4) + 5{,}000(P/F, 10\%, 4) - 25{,}000 = ?$

$6{,}620(3.170) + 5{,}000(.683) - 25{,}000 = -\$600 < 0 \therefore$

Do not purchase equipment

9-5 **

Agra Company is considering the construction of a certain facility. The facility will cost $1,000,000 to build. Its life and salvage value are expected to be 25 years and $100,000, respectively. Agra estimates the net income will be $200,000 per year for the first 15 years, and to decrease thereafter at $20,000 per year.

Agra will use a Minimum Attractive Rate of Return of 12% after taxes, an income tax rate of 46%, a capital gain tax rate of 28%, and a 10-year recovery period.

Which is the more economically advantageous depreciation schedule to use for this investment, the Accelerated Cost Recovery System (ACRS) or Straight Line (SL) depreciation? Show how you reached your decision. For ACRS use 8%, 14%, 12%, 10%, 10%, 10%, 9%, 9%, 9% and 9%, for years 1 through 10, respectively.

****

By inspection we would expect that ACRS depreciation would be the preferred method. But the problem asks for some proof.

To answer the question we do not need to perform a complete after-tax analysis, including all the cash flow information. We only need to consider the tax savings using either depreciation schedule. The Net Present Value (Worth) of each would be

$$NPV_{ACRS} = .46[(8)(P/F,.12,1) + 14(P/F,.12,2) + 12(P/F,.12,3) + 10(P/A,.12,3)(P/F,.12,2) + 9(P/A,.12,4)(P/F,.12,6)]$$

$$= \$26.58$$

$$NPV_{SL} = .46(10)(P/A,.12,10) = \$25.99$$

The ACRS percentages would be more economical.

9-6 **

The Tax Act of 1982 permits taxpayers who claim the Investment Tax Credit (ITC) to either (1) reduce the ITC fraction from 0.10 to 0.08 and depreciate the full first cost or (2) claim the full ITC of 0.10 but depreciate only the first cost less one-half the ITC (dollar amount) claimed. If a taxpayer, whose Minimum Attractive Rate of Return is 0.10, was considering the purchase of a qualitying asset with a 5-year recovery period, what is the maximum income tax rate that would make him indifferent between the two ITC choices above? Use either straight-line depreciation or the Accelerated Cost Recovery System (ACRS) percentages, which are 1, 15%; 2, 22%; 3, 21%; 4, 21%; 5, 21%.

****

One could solve the problem by assuming a hypothetical numerical example. The particular numbers would be irrelevant since they would cancel out. In the solution below FC represents the first cost of a qualifying asset and d the income tax rate. The solution uses straight line depreciation. The answer for ACRS percentages is also shown.

t	choice 1 tx svgs	choice 2 tx svgs	choice 1-2 tax svgs
0	.08 FC	.10 FC	-.02 FC
1-5	.20(d)FC	.2(d)(FC-.5(.1)FC)	+.01dFC

NPV = 0 = -.02 FC + .01d FC (P/A, 10%, 5)

d = 0.528 if SL used

d = 0.533 if ACRS used

** **9-7**

A company is considering the purchase of automatic equipment to improve productivity. The equipment costs $40,000, has an estimated life of 10 years and zero salvage value, and qualifies for the investment tax credit (ITC). The company will elect an ITC of 8% and depreciate the full first cost. The company's state income tax rate is 0.10 and their federal income tax rate is 0.48.

The annual disbursements (before taxes) for operation and maintenance are $14,000 per year. If the company uses straight line depreciation, and a 5-year recovery period, what minimum annual receipts (before taxes) must the company have to justify the equipment? Their Minimum Attractive Rate of Return (MARR) is 15% after taxes.

****

The minimum annual after tax cash flow A required to earn 15% is NPV = 0 = -36,000 + A(P/A, 15%, 10) thus A = +7,173 for t = 1, ···, 10. Because A'(1-d) + dD = A where A' is the annual before tax cash flow and D the annual depreciation, specifically, d = .1 + (.9)(.48) = .532 and D = 40,000/5 = 8,000 for t = 1, ···, 5. So

A' = ((7,173 (P/A, 15%, 10) - .532 (8,000)(P/A, 15%, 5)) (A/P, 15%, 10)/(1 - .532)

A' = 9,252 and the minimum annual receipts before taxes R' must be A' = R' - 14,000, R' = 14,000 + 9,252
= 23,253

9-8 **

By purchasing a truck for $30,000, a large profitable company in the 50% income tax bracket was able to save $5000 during year 1, $4000 during year 2, $3000 during year 3, $2000 during year 4 and $1000 during year 5. The company depreciated the truck using the Sum-Of-Years-Digits depreciation method over its four year depreciable life, while assuming a zero salvage value for depreciation purposes. The company wants to sell the truck at the end of year 5. Assume that capital gains/losses are taxed/refunded at a rate of 28%, and assume that the truck qualifies for a 10% investment tax credit.

What resale value will yield a 12% after-tax rate of return for the company?

****

Depreciation: $SOYD = \frac{4(5)}{2} = 10$

YEAR	SOYD Depreciation
1	4/10 (30,000 − 0) = 12,000
2	3/10 (30,000 − 0) = 9,000
3	2/10 (30,000 − 0) = 6,000
4	1/10 (30,000 − 0) = 3,000

ATCF Calculations:

YEAR	BTCF	SOYD depr.	TAXABLE INCOME	50% TAX	ATCF
0	-30,000	–	10% investment tax credit	+3,000	-27,000
1	5,000	12,000	-7,000	+3,500	8,500
2	4,000	9,000	-5,000	+2,500	6,500
3	3,000	6,000	-3,000	+1,500	4,500
4	2,000	3,000	-1,000	+500	2,500
5	1,000	–	1,000	-500	500
	R	–	capital gains of R taxed at 28%	-.28R	.72R

Solve for R:

$27{,}000 = 8{,}500(P/A, 12\%, 4) - 2{,}000(P/G, 12\%, 4) + (500 + .72R)(P/F, 12\%, 5)$

$27{,}000 = 8{,}500(3.037) - 2{,}000(4.127) + (500 + .72R)(.5674)$

$27{,}000 = 17{,}844.2 + .408528R$

$R = \frac{9{,}155.8}{.408528} = \underline{22{,}441.68}$

** **9-9**

As an investment, a large profitable corporation, with an incremental tax rate of 46%, purchased a tract of land with several buildings for $1,000,000. All the buildings but one, valued at $40,000, were demolished. This building was rented for storage. Straight line depreciation was used with a depreciable life of 10 years. However at the end of 5 years, the building was demolished and the land sold for $2,000,000 to a land developer. During the five years, no taxable income was attributed to the property because the rental income exactly equalled the property taxes and depreciation. The corporate capital gains tax rate is 28%. What was the after-tax rate of return?

****

Year	BTCF	Property Taxes	SL Deprec	Tax.Inc.	Taxes	ATCF
0	-1,000K		-	-	-	-1,000K
1-5	4K+x	x	4,000	0	0	4,000
S	2,000K		-	-	282K	1,718K

Land:

Initial Land Value = 1,000,000 - 40,000 = $960,000

Final Land Value = 2,000,000

Capital Gain = 2,000,000 - 960,000 = 1,040,000

Taxes on Gain = 0.28 x 1,040,000 = $291,200

Building:

Book Value = 40,000 - 5 x 4,000 = $20,000

Sale Price = 0 (Building demolished)

Ordinary Loss = Book Value - Sale Price = $20,000

Tax Savings = 0.46 x 20,000 = 9,200

Total Taxes = 291,200 - 9,200 = $282,000

NPW = 4,000 (P/A, i, 5) + 1,718,000 (P/F, i, 5) - 1,000,000 = 0

For i = 12%   3.605   0.5674   = -10,787

For i = 10%   3.791   0.6209   = 81,870

$$i = 10 + 2\left(\frac{81{,}870}{81{,}870 + 10{,}787}\right) = \underline{11.8\%}$$

9-10 **

A large company must build a bridge to have access to land for expansion of its manufacturing plant. The bridge could be fabricated of normal steel for an initial cost of $30,000 and should last 15 years. Maintenance (cleaning and painting) will cost $1000/year. If a more corrosion resistant steel were used, the annual maintenance cost would be only $100/year, although its life would be the same. In 15 years there will be no salvage for either bridge. The company pays an income tax rate of 48% and uses straight line depreciation. Ignore the investment tax credit.

If the minimum acceptable after tax rate of return is 12%, what is the maximum amount that should be spent on the corrosion resistant bridge?

****

Steel (A)

Year	BTCF	Deprec.	Tax. Inc.	Taxes@48%	ATCF
0	-30,000	—	—	—	-30,000
1-15	-1,000	2,000	-3,000	-1,440	+440
S	0	—	0	0	0

Resistant Steel (B)

Year	BTCF	Deprec.	Tax. Inc.	Taxes@48%	ATCF
0	-P	—	—	—	-P
1-15	-100	P/15	-100-P/15	(-48-0.032P)	-52+0.032P
S	0	-	0	0	0

$NPW_A = NPW_B$ for breakeven

$440(P/A, 12\%, 15) - 30{,}000 = (-52 + 0.032P)(P/A, 12\%, 15) - P$

$440(6.811) - 30{,}000 = (-52 + 0.032P)(6.811) - P$

$-27{,}003 = -354 + 0.218P - P$

$P = \frac{27{,}003 - 354}{1 - 0.218} = \underline{\$34{,}078}$

** **9-11**

A company has purchased a major piece of equipment which has a useful life of 20 years. An analyst is trying to decide on a maintenance program and has narrowed the alternatives to two. Alternative A is to perform $1000 of major maintenance every year. Alternative B is to perform $5000 of major maintenance only every fourth year. In either case, maintenance will be performed during the last year so that the equipment can be sold for an estimated $10,000. If the Minimum Attractive Rate of Return (MARR) is 18%, which maintenance plan should be chosen?

The analyst computed the solution as:

Equiv Uniform Annual Cost_A = $1000
Equiv Uniform Annual Cost_B = $5000(A/F, 18%, 4) = $958.5
Therefore choose Alternative B.

Is it possible the decision would change if income taxes were considered? Why or why not?

****

No. Both cash flows, which distinguish between the alternatives, would be reduced by the same percentage (i.e., taxes) so $EUAC_A > EUAC_B$ would still be true. If we assume a 45% tax rate, for example, the computations are as follows:

Alternative A -

Year	Before Tax Cash Flow	Taxable Income	Income Taxes	After Tax Cash Flow
1-4	-1,000	-1,000	+450*	-550

$EUAC_A = 550$

Alternative B -

Year	Befor Tax Cash Flow	Taxable Income	Income Taxes	After Tax Cash Flow
1-3	0	0	0	0
4	-5,000	-5,000	+2,250*	-2,750

$EUAC_B = 2{,}750\,(A/F, 18\%, 4) = 527.2$

* where + sign indicates a decrease in income taxes.

9-12 **

An unmarried student earned $5200 during the year. The student claims one $1000 exemption and estimates itemized deductions of $2600. If single people are automatically allowed $2300 of deducations, what is the student's Federal Income Tax?

Tax Rates:

Taxable Income		Tax is		
Over	But not over	This	Plus percentage	Over
$2300	$3400	$ 0	14%	$2300
3400	4400	154	16%	3400
4400	6500	314	18%	4400

****

Adjusted Gross Income	$5,200
Exemption	- 1,000
Itemized deductions (2,600 - 2,300)	- 300
Taxable Income	$ 3,900

Tax:

154 + 16% [3,900 - 3,400]

= 154 + .16 (500) = $ 234.

9-13 **

An asset with a five year ACRS* life will be purchased for $10,000. It will produce net annual benefits of $2000 per year for five years, after which it will have a net salvage value of zero and will be retired. The company's incremental tax rate is 46%. Ignore the investment tax credit and inflation. Calculate the after tax cash flows.

*The annual percentages to use are 15%, 22%, 21%, 21%, and 21% for years 1 through 5.

****

EOY	BTCF	5-YR ACRS	TAXABLE INCOME	CASH FLOW FOR TAXES	AFTER TAX CASH FLOW
0	-10,000				-10,000
1	2,000	1,500	500	-230	1,770
2	2,000	2,200	-200	92	2,092
3	2,000	2,100	-100	46	2,046
4	2,000	2,100	-100	46	2,046
5	2,000	2,100	-100	46	2,046

** **9-14**

Five years ago a lawyer bought an antique car for $50,000 as an investment. He has just sold the car for $70,000. The lawyer is single and has a taxable income of $50,000 from all income sources other than this investment. Antiques can not be depreciated. A portion of the appropriate tax table is as follows:

If Your Taxable Income Is		Your Tax Is		
Over	But Not Over	This	Plus Following Percentage	Over This
$41,500	$55,300	$10,319	42%	$41,500
55,300	81,800	16,115	48%	55,300
81,800		28,835	50%	81,800

(a) How much profit did he make on the sale after the required taxes have been paid?
(b) What was the after tax rate of return on the investment?

****

a) Incremental Tax Rate = 42% of next $5,300, then 48%

Capital Gain = Sale Price - Book Value = 70,000 - 50,000 = $20,000

Taxable Income = 40% of Capital Gain = (0.4)(20,000) = $8,000

Taxes = (0.42)(5,300) + (0.48)(8,000 - 5,300) = $3,522

After Tax Profit = (70,000 - 3,522) - 50,000 = <u>$16,478</u>

b)

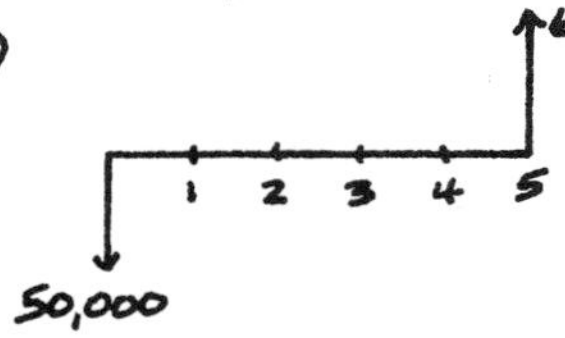

$(F/P, i, 5) = \frac{66,478}{50,000} = 1.3296$

$i = 5 + (1)\left(\frac{1.3296 - 1.276}{1.338 - 1.276}\right) = \underline{5.86\%}$

** **9-15**

A state tax of 10% is deductible from the income taxed by the Federal Government (Internal Revenue Service). The Federal tax is 40%. What is the combined effective tax rate?

****

$E.T. = S + (1 - S)f$

$= .1 + (1 - .1).4$

$= .1 + (.9)(.4) = .1 + .36 = .46 = \underline{46\%}$

9-16 **

A small oil company, in the 50% income tax bracket, bought an oil well for $1,000,000. Operating expenses (including labor costs, maintenance, utilities, etc.) are $500,000 per year plus $5 per barrel of oil pumped. A geologist calculated that the well contains 900,000 barrels of very low grade oil. The company expects to pump 300,000 barrels in year 1, 200,000 barrels in year 2, and 100,000 barrels in year 3. (It is not known yet what the pumping schedule will be for any year beyond year 3.) The company expects to sell the oil for $15 per barrel.

The company also purchased some oil pumping equipment for $2,000,000 and plans to depreciate the equipment using Accelerated Cost Recovery System (ACRS). This equipment is in the five year property class (applicable percentages are 20%, 32%, 24%, 16% and 8%, respectively, for years 1, 2, 3, 4 and 5). The company expects that the equipment will have no value at the end of its depreciable life.

Assuming the oil percentage depletion allowance is 15% and that this oil well has a zero salvage value, calculate the cost and percentage depletion allowances and the after-tax cash flows for years 1, 2, and 3 only. For each year be sure to use the depletion method that yields the largest deduction for that year. Fill in the table below.

Year	Percentage Depletion	Cost Depletion	After-Tax Cash Flow
1			
2			
3			

****

YEAR	1	2	3
Bbl	300K	200K	100K
Income - $15/Bbl	$4,500K	$3,000K	$1,500K
Expenses - $500 + $5/Bbl	$2,000K	$1,500K	$1,000K
ACRS depr.[1]	$400K	$640K	$480K
Before Depletion Taxable Inc.	$2,100K	$860K	$20K
15% Income	$675K	$450K	$225K
50% Limit	$1,050K	$430K	$10K
Allowable % depletion	$675K	$430K	$10K
Cost depletion allowance[3]	$333.$\bar{3}$K	$222.$\bar{2}$K	$111.$\bar{1}$K
Depletion allowance used[2]	$675K	430K	111.$\bar{1}$K

YEAR	1	2	3
Taxable Inc.	\$1,425K	\$430K	\$-91,111.11
50% Tax	\$712.5K	\$215.K	\$-45,555

YEAR	INCOME	EXPENSES	TAX	ATCF
1	\$4,500	\$2000K	\$712.5K	\$1,787.5K
2	3,000	1,500K	215K	1,285K
3	1,500	1,000K	-45,555	545,555

YEAR	% DEPLETION	COST DEPLETION	ATCF
1	\$675K	\$333.3̄3K	\$1,787.5K
2	\$430K	\$222.2̄2K	\$1,285K
3	\$10K	\$111.1̄1K	545,555

1 - ACRS = (\$2,000,000)(applicable %)

2 - depletion allowance used = 15% of Income

3 - Cost depletion allowance:

$$YR.1 = \frac{300K}{900K}(1{,}000K) = \$333.\bar{3}K$$

$$YR.2 = \frac{200K}{900K}(1{,}000K) = \$222.\bar{2}K$$

$$YR.3 = \frac{100K}{900K}(1{,}000K) = \$111.\bar{1}K$$

note: cost depletion allowance is compared on a year-by-year basis to the % depletion allowance to determine which method of calculating depletion will be used.

** **9-17**

For engineering economic analysis a corporation uses an incremental state income tax rate of 7.4% and an incremental federal rate of 46%. Calculate the effective tax rate.

****

Effective rate = 0.074 + (1 - 0.074)(0.46) = 0.5000

9-18 **

A company bought an asset at the begining of 1984 for $100,000. The company now has an offer to sell the asset for $60,000 at the end of 1985. For each of the depreciation methods shown below, determine the capital gain or loss that would be realized for 1985.

Depreciation Method	Depreciable Life	Salvage Value*	Capital Gain	Capital Loss
Straight Line	10 years	$ 1,000		
Sum-Of-Years Digits	5	25,000		
Double Declining Balance	4	0		
150% Declining Balance	15	0		

*This was assumed for depreciation purposes

****

Depreciation Method	Depreciable Life	Salvage Value	Capital Gain	Capital Loss
SL	10 yrs.	$1,000		$20,200
SOYD	5 yrs.	25,000	$5,000	
DDB	4 yrs.	0	$35,000	
150%DB	15 yrs.	0		$21,000

SL:

$\text{depr} = \frac{1}{10}(100{,}000 - 1{,}000) = 9{,}900$

$\text{book value} = 100{,}000 - 2(9{,}900) = 80{,}200$

$\text{cap. loss} = 80{,}200 - 60{,}000 = \$20{,}200$

SOYD:

$(\text{depr. yr. }1) + (\text{depr. yr. }2) = \frac{5+4}{15}(100{,}000 - 25{,}000) = 45{,}000$

$\text{book value} = 100{,}000 - 45{,}000 = 55{,}000$

$\text{capital gain} = 60{,}000 - 55{,}000 = \$5{,}000$

DDB:

$\text{depr. yr. }1 = \frac{2}{4}(100{,}000)$; $\text{depr. yr. }2 = \frac{2}{4}(100{,}000 - 50000)$

$\text{total depr} = 75{,}000$

$\text{book value} = 100{,}000 - 75{,}000 = 25{,}000$

$\text{capital gains} = 60{,}000 - 25{,}000 = \$35{,}000$

150% DB:

$\text{depr. yr. }1 = \frac{1.5}{15}(100{,}000) = 10{,}000$; $\text{depr. yr. }2 = \frac{1.5}{15}(90{,}000)$

$\text{total depr} = 19{,}000$

book value = 100,000 - 19,000
capital loss = 81,000 - 60,000 = $21,000

** **9-19**

A corporation's tax rate is 50%. An outlay of $35,000 is being considered for a new asset. Estimated annual receipts are $20,000 and annual disbursements $10,000. The useful life of the asset is 5 years and it has no salvage value.

(a) What is the prospective rate of return (ROR) before income tax?

(b) What is the prospective rate of return (ROR) after taxes, assuming straight line depreciation for writing off the asset for tax purposes?

****

$$Dc(SL) = \frac{1}{N}(P-S) = \frac{35{,}000-0}{5} = 7{,}000/\text{yr.}$$

YR	CFBT	Dc	TAX. INC.	50% TAX	CFAT
0	-35,000				-35,000
1	+10,000	7,000	+3,000	-1,500	+85,000
2	+10,000	7,000	+3,000	-1,500	+85,000
3	+10,000	7,000	+3,000	-1,500	+85,000
4	+10,000	7,000	+3,000	-1,500	+85,000
5	+10,000	7,000	+3,000	-1,500	+85,000
		Σ = 35,000			

a) ROR (BT) PW(B) = PW(C)

$A(P/A, i, n) = C$

$(P/A, i, 5) = \frac{35{,}000}{10{,}000} = 3.5$

12% = 3.605
x = 3.500
15% = 3.352

$\frac{x}{3} = \frac{.105}{.253}$ $x = 1.25$

(ROR) i = 12% + 1.25% = 13.25%

b) ROR (AT)

$(P/A, i, 5) = \frac{35{,}000}{8{,}500} = 4.1176$

6% = 4.212
x = 4.1176
7% = 4.11

$\frac{x}{1} = \frac{.0944}{.102}$ $x = .92549$

(ROR) i = 6% + 0.9% = 6.9%

9-20 **

A young couple has just won a prize in a nationally advertised magazine sweepstakes contest. The couple may receive their prize money in one of two ways: either $100,000 now or $25,000 per year for five years (with the first year payment to be received now). The couple wants to make a decision depending on the after-federal-tax consequences of the two options.

Assume that the couple currently has a combined taxable income of $28,000 and that the couple plans to continue working (and maintaining this same taxable income and filing a joint return) regardless of which payment option they choose. For what range of values of the after-tax Minimum Attractive Rate of Return (MARR) is the $100,000 now option the preferred alternative? For what range of values of the after-tax MARR is the $25,000 for five years option preferred? A portion of a federal income tax table is given below.

Tax Rates - If You File A Joint Return

If your taxable income is		Your tax is		
Over	But not over	This	Plus following percentage	Over this
$ 24,600	$ 29,900	$ 3,465	25%	$ 24,600
29,900	35,200	4,790	28	29,900
35,200	45,800	6,274	33	35,200
45,800	60,000	9,772	38	45,800
60,000	85,600	15,168	42	60,000
85,600	109,400	25,920	45	85,600
109,400	162,400	36,630	49	109,400
162,400		62,600	50	162,400

****

$25,000 OPTION:

- income increased from $28,000 to $53,000 for years 0,1,2,3,4.
- taxes paided on $53,000 are
  $9,772 + 38%(53,000 - 45,800) = $12,508
- let x = BTCF prior to the prize money

For years 0-4 ; BTCF = x + 25,000 , Taxable Income = 53k

Taxes = 12,508 , ATCF = x + 12,492

$100,000 OPTION:

- year 0: taxable income will increase from $28,000 to $128,000
- taxable income for years 1 to 4 will remain at $28,000
- income tax on $128,000 =
  $36,630 + 49% (128,000 - 109,400) = $45,744
- income tax on $28,000 =
  $3,465 + 25% (28,000 - 24,600) = $4,315

For year 0; BTCF = x + 100,000, Taxable Income = 128,000
Taxes = 45,744, ATCF = x + 54,256

For years 1-4; BTCF = x, Taxable Income = 28,000
Taxes = 4,315, ATCF = x - 4,315

YEAR	ATCF (25K OPTION - 100K OPTION)
0	-41,764
1	+16,807
2	+16,807
3	+16,807
4	+16,807

41,764 = 16,807 (P/A, i%, 4)

(P/A, i%, 4) = 2.4849

i = 20%	2.589	by interpolation:
i = ?	2.4849	$i = 20\% + (25-20)\left(\frac{2.589 - 2.4849}{2.589 - 2.362}\right)$
i = 25%	2.362	

i = 22.293%

choose $100,000 option if MARR > 22.293%

choose $25,000 option if MARR ≤ 22.293%

9-21 **

A delivery firm is considering the purchase of a light truck. The purchase price is $11,500. The truck will be purchased January 1, 1986, if needed. The truck will be sold after three years for $2000.

The ACRS depreciation schedule (3-year class property) for equipment placed in service after 1985 is:

Year	Recovery Factor
1	33%
2	45
3	22

There is a 6% investment tax credit on 3-year property if it is held for three years. The firm has a combined state and federal income tax rate of 50%. Capital gains are taxed at the combined rate of 35%. Fill in the table below. If the firm has an after tax Minimum Attractive Rate of Return (MARR) of 10%, is the investment desirable?

Year	Before Tax Cash Flow	ACRS Depreciation	Taxable Income	Income Taxes	Investment Tax Credit	After Tax Cash Flow
0	-$11,500					
1	+4,000					
2	+6,000					
3	+5,000					
	+2,000(Salvage)					

****

Year	Before Tax Cash Flow	ACRS Depreciation	Taxable Income	Income Taxes	Investment Tax Credit	After Tax Cash Flow
0	-$11,500				+690	-10,810.00
1	+4,000	-3,795	+205	-102.50		+3,897.50
2	+6,000	-5,175	+825	-412.50		+5,587.50
3	+5,000	-2,530	+2,470	-1,235.00		+3,765.00
	+2,000(Salvage)		+2,000	-700		+1,300.00

a)

YR	DEPRECIATION
1	0.33 × 11,500 = 3,795
2	0.45 × 11,500 = 5,175
3	0.22 × 11,500 = 2,530

b) Salvage taxed at a capital gain (ACRS depreciates to 0 book value)

2,000 × 0.35 = 700 taxes

The combined are from 28% federal and 9.6% state

c) investment tax credit = 0.06 × 11,500 = +690.

Check desirability of investment:

$$NPW\ @\ 10\%\ \text{interest} = -10{,}810 + 3{,}897.50(P/F,10\%,1) + 5{,}587.50(P/F,10\%,2) + (3{,}765+1{,}300)(P/F,10\%,3)$$

$$= +1{,}156.06 > 0 \quad \therefore \text{Project is acceptable}$$

** **9-22**

A large and profitable company, in the 50% income tax bracket, is considering the purchase of a new piece of machinery that qualifies for a 10% investment tax credit. The new machine will yield benefits of \$10,000 for year 1, \$15,000 for year 2, \$20,000 for year 3, \$20,000 for year 4, and \$20,000 for year 5.

The machinery is to be depreciated using the Accelerated Cost Recovery System (ACRS) over three years. The percentages are 33%, 45% and 22%, respectively, for years 1, 2 and 3. The company believes the machinery can be sold at the end of five years of use for 25% of the original purchase price.

What is the maximum purchase cost the company can pay if it requires a 12% after-tax rate of return? Assume, if necessary, that capital gains/losses are taxed/refunded at a rate of 28%.

****

YEAR	BTCF	ACRS depr	TAXABLE INCOME	50% TAX	ATCF
0	-P	—	investment tax credit	+.10P	-.90P
1	10,000	.33P	10,000-.33P	5,000+.165P	5,000+.165P
2	15,000	.45P	15,000-.45P	7,500+.225P	7,500+.225P
3	20,000	.22P	20,000-.22P	10,000+.110P	10,000+.110P
4	20,000	0	20,000	10,000	10,000
5	{ 20,000 .25P	0	20,000 .25P capital gain taxed at 28%	10,000 .07P	{ 10,000 .18P

$$.90P = 5{,}000(P/A,12\%,5) + 2500(P/G,12\%,3) + 5{,}000(P/A,12\%,2)(P/F,12\%,3) + .18P(P/F,12\%,5)$$

$$= 5{,}000(3.605) + 2{,}500(2.221) + 5{,}000(1.690)(.7118) + .18P(.5674)$$

$$.90P = 29{,}592.21 + .102132P \rightarrow \underline{P = 37{,}089.10}$$

9-23

An investment of $120,000 is being spent on modifications to the pasteurization, packaging lines and distribution vans at a large dairy to increase its productivity. Of the $120,000 total investment, $85,000 corresponds to equipment depreciable only by straight line with a salvage value of $5000 after 8 years. The remaining $35,000 are depreciated by double declining balance with conversion to straight line (at the optimum point), with a salvage value of $3000 after 8 years. The benefits per year will amount to $32,000 for the next 8 years. If this dairy has an incremental tax rate of 48%, what is the payback period before and after taxes for this investment? Assume that at the end of the 8 years, the depreciable equipment is sold at its exact salvage value.

1. The depreciation schedules:

a) P = $85,000 S = $5,000 n = 8 ; straight line
depreciation charge (any yr.) = $\frac{1}{N}(P-S) = \frac{1}{8}(85K-5K) = 10K$

b) P = $35,000 S = $3,000 n = 8 { straight line (SL) / double declining balance (DDB) }

For SL depr = $\frac{\text{book value at beginning of year} - S}{\text{remaining useful life at beginning of year}}$

For DDB depr. = $\frac{2}{N}(\text{Book Value}) = \frac{2}{8}(\text{Book Value})$

YR.	DDB	SL	Choose
1	$\frac{2}{8}(35K) = 8{,}750$	$\frac{1}{8}(35K-3K) = 4{,}000$	DDB
2	$\frac{2}{8}(35K-8{,}750) = 6{,}562.5$	$\frac{1}{7}(35K-8{,}750-3K) = 3{,}321.4$	DDB
3	$\frac{2}{8}(35K-15{,}312.5) = 4{,}921.87$	$\frac{1}{6}(19{,}687.5-3K) = 2{,}781.25$	DDB
4	$\frac{2}{8}(35K-20{,}234.37) = 3{,}691.4$	$\frac{1}{5}(14{,}765.63-3K) = 2{,}353.13$	DDB
5	$\frac{2}{8}(35K-23{,}925.77) = 2{,}768.55$	$\frac{1}{4}(11{,}074.23-3K) = 2{,}018.56$	DDB
6	$\frac{2}{8}(35K-26{,}694.32) = 2{,}076.42$	$\frac{1}{3}(8{,}305.67-3K) = 1{,}768.56$	DDB
7	$\frac{2}{8}(35K-28{,}770.74) = 1{,}557.31$	$\frac{1}{2}(6{,}229.25-3K) = 1{,}614.63$	SL
8	————	1,614.63	SL

2. The cash flows (before and after taxes)

YR.	Before Tax Cash Flow	Dep. [total of (a)+(b)]	Δ Taxable Income	Taxes @ 48%	After-tax Cash Flow	Cummulative After-tax Benefits
0	-120K				-120K	0
1	32K	18,750	13,250	6,360	25,640	25,640
2	32K	16,562.5	15,437.5	7,410	24,590	50,230
3	32K	14,921.87	17,078.13	8,197.5	23,802.5	74,032.5
4	32K	13,691.4	18,303.6	8,788.13	23,211.87	97,244.37

YR.	Before Tax Cash Flow	Dep. [total of (a)+(b)]	Δ Taxable Income	Taxes @ 48%	After-tax Cash Flow	Cummulative After-tax Benefits
5	32K	12,768.55	19,231.45	9,231.09	22,768.91	120,013.28**
6	32K	12,076.42	19,923.58	9,563.32	22,436.68	
7	32K	11,614.63	20,385.37	9,784.97	22,215.03	
8	{32K, 8K	11,614.63	{20,385.37, 8,000	{9,784.97, 0*	30,215.03	

* no capital gain/loss
** payback

The payback period before taxes is $\frac{120{,}000}{32{,}000} = 3.75$ years
The payback period after taxes is approx. 5 years

** **9-24**

A large and profitable company in the 50% income tax bracket bought an asset for $100,000. The asset will be depreciated over 10 years using 175% Declining Balance depreciation with a $75,000 salvage value assumed. The asset is expected to yield benefits of $15,000 per year. Calculate the After Tax Cash Flow for years 1, 2 and 3 only.

****

YEAR	BTCF	175% DB	TAXABLE INCOME	50% TAX	ATCF
0	−100,000	−	−	−	−100,000
1	+15,000	17,500[1]	−2,500	+1,250	+16,250
2	+15,000	7,500[2]	+7,500	−3,750	+11,250
3	+15,000	0	+15,000	−7,500	+7,500

1 - Book value at end of year 1 $= 100{,}000 - \frac{1.75}{10}(100{,}000) = 82{,}500$
∴ BOOK VALUE = 100,000 − 17,500 = 82,500
2 - at end of year 2 : $\frac{1.75}{10}(82{,}500) = 14{,}437.5$
∴ BOOK VALUE = 100,000 − 17,500 − 14,437.5 = 68,062.5
since this value is less than the salvage value the asset may only be depreciated by 82,500 − 75,000.

9-25 **

A manufacturing firm purchases a machine in January for $100,000. The machine has an estimated useful life of 5 years, with an estimated salvage value of $20,000. The use of the machine should generate $40,000 before-tax profit each year over its 5-year useful life.

PART 1. Complete the following table.

Year	Before Tax Cash Flow	Sum of Digits Depreciation	Taxable Income	Taxes at 40%	After Tax Cash Flow
0					
1					
2					
3					
4					
5					

PART 2. Does the sum of digits depreciation represent a cash flow?

PART 3. Calculate the Before-Tax Rate of Return and the After-Tax Rate of Return.

****

PART 1:

Before Tax Cash Flow -

In January you must pay \$100,000. At the end of the first year, and in each subsequent year you realize a cash income of \$40,000. This income less depreciation allowance is taxable. In year 5 you also realize \$20,000 from salvage of the equipment. This amount is not taxable as it represents a capital expense that was never allocated as depreciation.

Sum of Years Digits Depreciation -

year 1	= 5/15 (\$100,000 - \$20,000) =	\$26,667
year 2	= 4/15 (\$80,000) =	\$21,333
year 3	= 3/15 (\$80,000) =	\$16,000
year 4	= 2/15 (\$80,000) =	\$10,667
year 5	= 1/15 (\$80,000) =	\$5,333
check		80,000

Taxable Income -

year 1 = 40,000 - 26,667 = $13,333

year 2 = 40,000 - 21,333 = $18,667

year 3 = 40,000 - 16,000 = $24,000

year 4 = 40,000 - 10,667 = $29,333

year 5 = 40,000 - 5,333 = $34,667

Taxes at 40% -

year 1 = $13,333 x .4 = $5,333

year 2 = $18,667 x .4 = $7,467

year 3 = $24,000 x .4 = $9,600

year 4 = $29,333 x .4 = $11,733

year 5 = $34,667 x .4 = $13,867

After Tax Cash Flow -

year 0 = -100,000

year 1 = 40,000 - 5,333 = $34,667

year 2 = 40,000 - 7,467 = $32,533

year 3 = 40,000 - 9,600 = $30,400

year 4 = 40,000 - 11,733 = $28,267

year 5 = 60,000 - 13,867 = $46,133

Year	Before Tax Cash Flow	Sum of Digits Depreciation	Taxable Income	Taxes at 40%	After Tax Cash Flow
0	-100,000	—	—	—	-100,000
1	40,000	26,667	13,333	5,333	34,667
2	40,000	21,333	18,667	7,467	32,533
3	40,000	16,000	24,000	9,600	30,400
4	40,000	10,667	29,333	11,733	28,267
5	40,000 20,000	5,333	34,667	13,867	46,133

PART 2:

The sum of digits depreciation is a book keeping allocation of capital expense for purposes of computing taxable income. In itself it does not represent a cash flow.

PART 3:

Before Tax Rate of Return –

$100{,}000 = 40{,}000\,(P/A, i, 5) + 20{,}000\,(P/F, i, 5)$

at 35% $100{,}000 \neq 40{,}000\,(2.220) + 20{,}000\,(.2230) \neq 93.26$

at 30% $100{,}000 \neq 40{,}000\,(2.436) + 20{,}000\,(.2693) \neq 102.83$

Interpolating $\dfrac{102.83-100}{102.83-93.26} = \dfrac{\Delta i}{5\%}$

$\Delta i = 1.48\%$ or $i = 30\% + 1.5\% = 31.5\%$

After Tax Rate of Return –

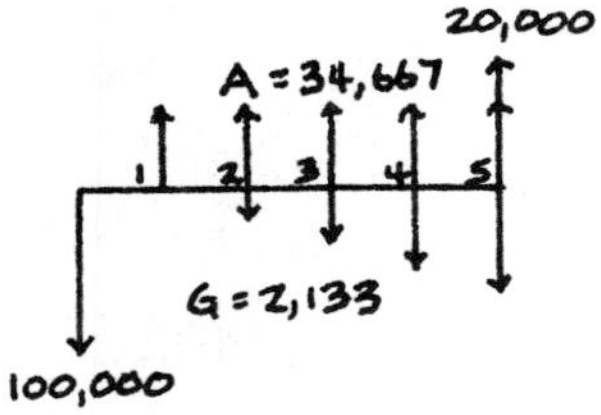

$100{,}000 = 34{,}667\,(P/A, i\%, n) - 2{,}133\,(P/G, i\%, n) + 20{,}000\,(P/F, i\%, n)$

at 20%: $100{,}000 \neq 34{,}667\,(2.991) - 2{,}133\,(4.906)$
$+ 20{,}000\,(.4019) \neq 101{,}263$

at 25%: $100{,}000 \neq 34{,}667\,(2.689) - 2{,}133\,(4.204)$
$+ 20{,}000\,(.3277) \neq 90{,}807$

Interpolating: $\dfrac{101{,}263 - 100{,}000}{101{,}263 - 90{,}807} = \dfrac{\Delta i}{5\%}$

$\Delta i = .60\%$ and $i = 20\% + .60 = 20.60\%$

** **9-26**

PARC, a large profitable firm, has an opportunity to expand one of its production facilities at a cost of $375,000. The equipment is expected to have an economic life of 10 years and to have a resale value of $25,000 after 10 years of use. If the expansion is undertaken, PARC expects that their income will increase by $60,000 for year 1, and then increase by $5000 each year through year 10 ($65,000 for year 2, $70,000 for year 3, ..., $105,000 for year 10). If the equipment is purchased, PARC will depreciate it using 7-year Double Declining Balance depreciation changing to Straight Line, with zero salvage value at the end of year 7 for tax purposes. PARC will take a 10% investment tax credit if the equipment is purchased.

The annual operating cost is expected to be $5000 for the first year and to increase by 5% per year ($5250 for year 2, $5512.50 for year 3, ..., $7756.64 for year 10). If the equipment is purchased, PARC will pay $175,000 down and finance the balance with a 5-year 12% loan payable in annual payments (the annual payments will be rounded to the closest $500 amount), which include both principal and interest. Since PARC is a "large and profitable" firm their tax rate is 46% and their capital gains tax rate is 28%.

If PARC's Minimum Attractive Rate of Return (MARR) is 15%, should they undertake this expansion?

****

Depreciation

$Year_t$	$Book\ Value_t$	$Deprec_t$ DDB	$Deprec_t$ Str. Line
0	375,000		
1	267,857.14	107,142.86	~~53,571.43~~
2	191,326.53	76,530.61	~~44,642.86~~
3	136,661.81	54,664.72	~~38,265.31~~
4	97,615.58	39,046.23	~~34,165.45~~
5	65,077.05	~~27,830.17~~	32,538.53
6	32,538.53	–	32,538.53
7	0	–	32,538.53

loan 200,000 5yr 12%

$$A = 200{,}000 \times \frac{.12}{1-(1.12)^{-5}} = 55{,}481.95 \rightarrow 55{,}500.$$

Five year 12% Loan :

Year	Balance	Interest	Total Payment	Debt Reduction
0	200,000	24,000	55,500	31,500
1	168,500	20,220	55,500	35,280
2	133,220	15,986.4	55,500	39,513.6
3	93,706.4	11,244.77	55,500	44,255.23
4	49,451.17	5,934.14	55,385.31	49,451.17
5	0	–	–	–

Year	Income	Expense	Net
0	0	175,000	-175,000
1	60,000	5,000	55,000
2	65,000	5,250	59,750
3	70,000	5,512.5	64,487.5
4	75,000	5,788.13	69,211.87
5	80,000	6,077.53	73,922.47
6	85,000	6,381.41	78,618.59
7	90,000	6,700.48	83,299.52
8	95,000	7,035.50	87,964.5
9	100,000	7,387.28	92,612.72
10	105,000 25,000 capital gain	7,756.64	97,243.36

YEAR	NET OPERATING PROFIT	INTEREST	DEPRECIATION	TAXABLE INCOME	TAXES (46%)	LOAN PAYMENTS	AFTER TAX CASH FLOW	PRESENT WORTH AFTER TAX CASH FLOW
0	-175,000	—	—	-375,000 FOR IC	-37,500 ITC	—	-137,500	-137,500
1	55,000	24,000	107,142.86	- 76,142.86	-35,025.72	55,500	34,525.72	30,022.36
2	59,750	20,220	76,530.16	- 37,000.61	-17,020.28	55,500	21,270.28	16,083.39
3	64,487.5	15,986.4	54,664.72	- 6,163.62	-2,835.27	55,500	11,822.77	7,773.66
4	69,211.87	11,244.77	39,046.23	18,920.87	8,703.60	55,500	5,008.27	2,863.49
5	73,922.47	5,934.14	32,538.53	35,449.8	16,306.91	55,385.31	2,230.25	1,108.83
6	78,618.59	—	32,538.53	46,080.06	21,196.83	—	57,421.76	24,825.01
7	83,299.52	—	32,538.53 (0 SV)	50,760.99	23,350.06	—	59,949.47	22,537.23
8	87,904.5	—	—	87,964.5	40,463.67	—	47,500.83	15,528.11
9	92,612.72	—	—	92,612.72	42,601.85	—	50,010.87	14,216.21
10	97,243.36	—	—	97,243.36	44,731.95	—	52,511.41	12,980.02
	25,000 CG*			25,000 CG @ 28%	7,000		18,000	4,449.32
								+14,887.64

* CG: Capital gains

Conclusion: Expansion is desirable

9-27 **

A married man with a taxable income of $45,000 buys a house to live in on January 1 for $80,000, putting $8000 down and financing the remainder at 12% for 30 years, in 360 monthly payments, payable at the end of each month. Taxes on the house are $400 per year, payable at the end of the year. Hazard insurance of $150/year is payable at the beginning of the year. Maintenance costs are $500/year, payable at the end of the year.

(a) What is the monthly payment on the loan?
(b) If he can deduct interest paid on the loan and property taxes from his taxable income (assume a 33% incremental tax rate), what is his effective monthly payment during the first year?
(c) If he sells the house 3 years later on January 1 for $130,000, what additional income tax will he pay that year due to the sale if he does not reinvest that money? Assume that 40% of the long term capital gain is taxed at a 33% incremental tax rate.

****

Exact solution using hand calculator

a) $A = (80{,}000 - 8{,}000)(A/P, 1\%, 360) = 72{,}000(0.010286) = \underline{740.60}$

b) Amount owed after one year

$P = 740.60(P/A, 1\%, 348) = 740.60(96.8655) = 71{,}739$

∴ Total principle payment in first year = 72,000 − 71,739 = $261

∴ Total interest payment in first year = 12(740.60) − 261 = $8,626

Decrease in taxable income = 8,626 + 400 = 9,026

Decrease in income taxes for first year = 9,026 × 0.33 = 2,979

∴ Effective monthly payment $= 740.60 - \frac{2{,}979}{12} = \underline{\$492.35}$

c) Since the dwelling was used as the man's personal home, he could not deduct depreciation while he owned it.

Long Term capital gain = 130,000 − 80,000 = $50,000

Tax on capital gain = 0.40(50,000)(0.33) = $\underline{\$6{,}600}$

9-28 **

A young man bought a one-year savings certificate for $10,000, which pays 15%. Before he buys the certificate he has a taxable income that puts him in the 34% incremental income tax rate. What is his after taxes rate of return on his investment?

****

At ROR = (1 − Δtax rate)(BT ROR) = (1 − .34)(.15) = $\underline{9.9\%}$

** **9-29**

A project can be summarized by the data given in the table below. The company uses straight line depreciation, pays an incremental income tax rate of 30% and requires an after-tax rate of return of 12%. Ignore any investment tax credit.

Item	Amount
Life	30 years
Investment	$75,000
Salvage	15,000
Revenues	26,000/year
Operating Expenses	13,500/year

(a) Using Net Present Worth, determine whether the project should be undertaken.

(b) If the company used Sum-Of-Years-Digits depreciation, is it possible the decision would change? (No computations needed.)

****

a) $\text{Depreciation} = \dfrac{75{,}000 - 15{,}000}{30} = 2{,}000/\text{yr.}$

Year	BTCF	Depreciation	Taxable Income	Taxes @ 30%	ATCF
0	-75,000	–	–	–	-75,000
1-30	12,500	2,000	10,500	3,150	9,350
S	15,000	–	–	–	15,000

$$NPW = 9{,}350(P/A, 12\%, 30) + 15{,}000(P/F, 12\%, 30) - 75{,}000$$
$$= \underline{+815}$$

Yes, take project since NPW > 0

b) No. Although total depreciation is the same, SOYD is larger in the early years when it is worth more. Therefore the NPW would <u>increase</u> with SOYD making the project even more desirable.

9-30 ***

A corporation expects to receive $32,000 each year for 15 years from the sale of a product. There will be an initial investment of $150,000. The expenses of manufacturing and selling the product will be $7530 per year. Assume straight line depreciation, a 15-year useful life and no salvage value. Use a 48% income tax rate and ignore the investment tax credit.

Determine the projected after-tax rate of return.

****

Straight line depreciation $= \frac{P-F}{n} = \frac{150{,}000-0}{15} = \$10{,}000$ per year

Year	Before Tax Cash Flow	Straight Line Depreciation	Taxable Income	48% Income Taxes	After Tax Cash Flow
0	-150,000				-150,000
1-15	+24,470	10,000	14,470	6,946	+17,524

Take the After Tax Cash Flow and compute the rate of return at which PW of Costs equals PW of Benefits.

$150{,}000 = 17{,}524\,(P/A, i\%, 15)$

$(P/A, i\%, 15) = \frac{150{,}000}{17{,}524} = 8.559$

From interest tables, $i = 8\%$

****

A NOTE ON THE RECAPTURE OF DEPRECIATION

****

When property is sold for more or less than its book value, engineering economy textbooks often treat the difference as a capital gain or loss. This may or may not be accurate.

When the book value of the property has been reduced by depreciation, U.S. tax laws may treat some or all of the gain or loss as ordinary income or ordinary loss. The recapture of depreciation rules are complex. This may explain why textbooks avoid discussing the problem in detail.

The important thing to recognize is that there may be less favorable tax treatment than suggested in textbooks. Further, of course, if you are solving problems in a college course you need to follow the professor's approach - since he grades the exams!

10
EQUIPMENT RETIREMENT AND REPLACEMENT

** 10-1

One of the four ovens at a bakery is being considered for replacement. Its salvage value and maintenance costs are given in the table below for several years. A new oven costs $80,000 and this price includes a complete guarantee of the maintenance costs for the first two years, and it covers a good proportion of the maintenance costs for years 3 and 4. The salvage value and maintenance costs are also summarized in the table.

	Old Oven		New Oven	
Year	Salvage value at end of year	Maintenance costs	Salvage value at end of year	Maintenance costs
0	$20,000	$ -	$80,000	$ -
1	17,000	9,500	75,000	0
2	14,000	9,600	70,000	0
3	11,000	9,700	66,000	1,000
4	7,000	9,800	62,000	3,000

Both the old and the new ovens have similar productivities and energy costs. Should the oven be replaced this year, if the MARR equals 10%?

****

1. The old oven ("defender")

YR.	S value at end-of-year	EUAC Capital Recovery =(P-S)×(A/P,10%,n)+Si	Maint. Cost	EUAC of Maint. =9,500+100(A/G,10%,n)	EUAC Total
0	P=20k	—	—	—	—
1	17K	5,000	9,500	9,500	14,500
2	14K	4,857.2	9,600	9,547.6	14,404.8
3	11K	4,718.9	9,700	9,593.7	14,312.6*
4	7K	4,801.5	9,800	9,638.1	14,439.6

* Economic Life = 3 years, with EUAC = 14,312.6

2. The new oven ("challenger")

YR.	S value at end-of-year	EUAC Capital Recovery = (P−S)×(A/P,10%,n)+Si	Maint. Cost	EUAC of Maint	EUAC Total
0	P = 80k	—	—	—	—
1	75K	13,000	0	0	13,000
2	70K	12,762	0	0	12,762
3	66K	12,229.4	1,000	302.1 (a)	12,531.5 **
4	62K	11,879	3,000	883.55 (b)	12,762.55

(a) $1000(A/F,10\%,3) = 302.1$

(b) $[1{,}000(F/P,10\%,1)+3{,}000](A/F,10\%,4) = 883.55$

** Economic Life 3 years, with EUAC of 12,531.5

Since EUAC defender > EUAC challenger (14,312.6 > 12,531.5) replace oven this year.

10-2 **

The cash flow diagram below indicates the costs associated with a piece of equipment. The investment cost is $5000 and there is no salvage. During the first 3 years the equipment is under warranty so there are no maintenance costs. Then the estimated maintenance costs over 15 years follow the pattern shown. To show you can do the calculations required to find the most economic useful life, determine the Equivalent Uniform Annual Cost (EUAC) for n = 12 if the Minimum Attractive Rate of Return (MARR) = 15%. You must use gradient and uniform series factors in your solution.

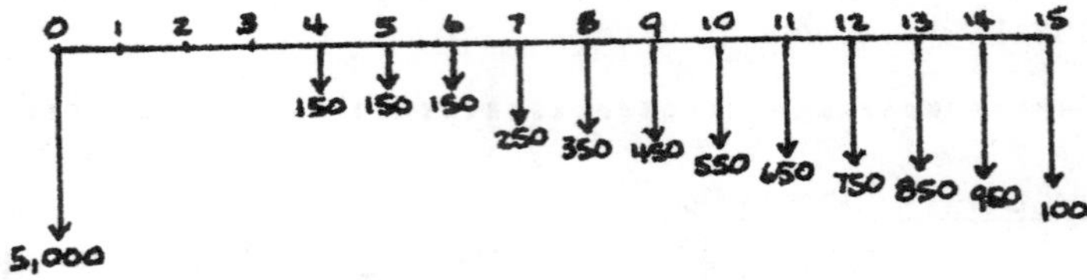

****

$$EUAC_{12} = 5{,}000(A/P,15\%,12) + 150(F/A,15\%,9)(A/F,15\%,12)$$
$$+ 100(P/G,15\%,7)(P/F,15\%,5)(A/P,15\%,12)$$
$$= 5{,}000(.1845) + 150(16.786)(.0345) + 100(10.192)(.4972)(.1845)$$
$$= \$\underline{1{,}103}$$

** **10-3**

A hospital is considering buying a new $40,000 diagnostic machine which will have no salvage value after installation as the cost of removal equals any sale value. Maintenance is estimated at $2000 per year as long as the machine is owned. After ten years the radioactive ion source will have caused sufficient damage to machine components that safe operation is no longer possible and the machine must be scrapped.
The most economic life of this machine is:
(Select one)
(a) One year since it will have no salvage after installation.
(b) Ten years because maintenance doesn't increase.
(c) Less than ten years but more information is needed to determine it.

****

The correct answer is (b).

** **10-4**

A petroleum company, whose Minimum Attractive Rate of Return is 10%, needs to paint the vessels and pipes in its refinery periodically to prevent rust. "Tuff-Coat," a durable paint, can be purchased for $8.05 a gallon while "Quick-Cover," a less durable paint, costs $3.25 a gallon. The labor cost of applying a gallon of paint is $6.00. Both paints are equally easy to apply and will cover the same area per gallon. Quick-Cover is expected to last 5 years. How long must Tuff-Cover promise to last to justify its use?

****

This replacement problem requires that we solve for a breakeven point. Let N represent the number of years Tuff-Cost must last. The easiest measure of worth to use in this situation is Equivalent Uniform Cash Flow (EUCF). Although more computationally cumbersome, others could be used and if applied correctly they would result in the same answer.

Find N such that $EUCF_{TC} = EUCF_{QC}$

$14.05(A/P,N,.1) = 9.25(A/P,5,.1)$

$(A/P,N,.1) = 0.17367$ therefore $N = 9$ years.

Tuff-coat must last at least 9 years. Notice that this solution implicitly assumes that the pipes need to be painted indefinitely (i.e., forever) and that the paint and costs of painting never change (ie. no inflation or technological improvements effecting the paint or the cost to produce and sell paint, or to apply the paint.)

10-5 **

Ten years ago Hyway Robbery, Inc. installed a conveyor system for $8000. The conveyor has been fully depreciated to its zero salvage value. The company is considering replacing the conveyor because maintenance costs have been increasing. The estimated end-of-year maintenance costs for the next five years are as follows:

Year	Maintenance
1	$1000
2	1250
3	1500
4	1750
5	2000

At any time the cost of removal just equals the value for scrap metal. The replacement the company is considering has an after-tax Equivalent Uniform Annual Cost (EUAC) of $858 at its most economic life. The company pays a 48% incremental income tax rate and requires a Minimum Attractive Rate of Return (MARR) of 10% after taxes.

(a) Should the conveyor be replaced now? Show the basis used for the decision.

(b) Now assume the old conveyor could be sold at any time as scrap metal for $500 more than the cost of removal. All other data remain the same. Should the conveyor be replaced?

****

a) Since the current value ($0.00) is not changing but maintenance costs are increasing, the most economic life is one year.

Year	BTCF	Tax. Inc.	Taxes	ATCF
0	0	0	0	0
1	-1,000	-1,000	-480	-520
S	0	0	0	0

Defender uniform equivalent cost: $EUAC_D = \$520$

Since $EUAC_D < EUAC_C$, Keep the old conveyor for now.

b)

Year	BTCF	Tax. Inc.	Taxes	ATCF
0	-500	—	—	-260*
1	-1,000	-1,000	-480	-520
S	+500	+500	+240	+260

* If sold now:

BTCF	Tax. Inc.	Taxes	ATCF	
+500	+500	+240	+260	1st cost is foregone salvage value.

Note: Whenever the conveyor is sold there is an ordinary gain of \$500 since the book value is zero.

$EUAC_D = 520 + 260(A/P, 10\%, 1) - 260(A/F, 10\%, 1) = \$546$

Since $EUAC_D < EUAC_C$, keep the old conveyor for now.

** **10-6**

Ten years ago, the Cool Chemical Company installed a heat exchanger in its plant for \$10,000. The company is considering replacing the heat exchanger because maintenance costs have been increasing. The estimated maintenance costs for the next 5 years are as follows:

Year	Maintenance
1	\$1000
2	1200
3	1400
4	1600
5	1800

Whenever the heat exchanger is replaced, the cost of removal will be \$1500 more than the heat exchanger is worth as scrap metal. The replacement the company is considering has an Equivalent Uniform Annual Cost (EUAC) = \$900 at its most economic life. Should the heat exchanger be replaced now if the company's Minimum Attractive Rate of Return (MARR) is 20%? Ignore taxes.

****

a) Since the current value (\$-1,500) is not changing but maintenance cost are increasing, the most economic life is one year.

b)

Year	Cash Flow	
0	+1,500	(Foregone Salvage)
1	-1,000	(Maintenance)
S	-1,500	(Negative Salvage)

c) Uniform equivalent cost of the defender:

$EUAC_D = 1{,}000 + 1{,}500(A/F, 20\%, 1) - 1{,}500(A/P, 20\%, 1) = \$700$

d) Since $EUAC_D < EUAC_C$, keep the old heat exchanger for now.

10-7 **

An engineer is trying to determine the economic life of a new metal press. The press costs $10,000 initially. First year maintenance cost is $1,000. Maintenance cost is forecast to increase $1,000 per year for each year after the first. Fill in the table below and determine the economic life of the press. Consider only maintenance and capital recovery in your analysis. Interest is 15%.

Year	Maintenance Cost	EUAC* of Capital Recovery	EUAC of Maintenance	Total EUAC
1	$1000			
2	2000			
3	3000			
4	4000			
5	5000			
6	6000			
7	7000			
8	8000			

*EUAC = Equivalent Uniform Annual Cost

****

Year	Maintenance Cost	EUAC of Capital Recovery	EUAC of Maintenance	Total EUAC
1	$1000	$11,500	1,000	12,500
2	2000	6,151	1,465	7,616
3	3000	4,380	1,907	6,287
4	4000	3,503	2,326	5,829
5	5000	2,983	2,723	5,706
6	6000	2,642	3,097	5,739
7	7000	2,404	3,450	5,854
8	8000	2,229	3,781	6,010

Economic Life = 5 yrs (EUAC = minimum)

EUAC of Capital Recovery = $10,000(A/P, 15%, n)
EUAC of Maintenance = $1,000 + 1,000(A/G, 15%, n)

*** **10-8**

A manufacturer is contemplating the purchase of an additional forklift truck to improve material handling in the plant. He is considering two popular models, the Convair T6 and the FMC 340. The relevant financial data are shown below. The manufacturer will use a 5-year recovery period and the Accelerated Cost Recovery System (ACRS) percentages: 1st yr, 15%; 2nd yr, 22%; 3rd-5th yr, 21% each year. The manufacturer's income tax rate is 50% and his Minimum Attractive Rate of Return (MARR) is 12% after taxes.

Model	First Cost	Life	Salvage Value	Annual Operating Expenses
Convair T6	$20,000	5 yrs	$2000	$8000
FMC 340	29,000	7	4000	4000

(a) Which model is more economical?
(b) List two important assumptions that are implicit in your computations in (a).

****

a)

Compute the EUCF for each model.

Convair:

$$\begin{aligned} EUCF &= 20{,}000(A/P,12\%,5) + 2{,}000(.5)(A/F,12\%,5) \\ &\quad - (.5)8{,}000 + [20{,}000(.15)(P/F,12\%,1) \\ &\quad + 20{,}000(.22)(P/F,12\%,2) + 20{,}000(.21) \\ &\quad (P/A,12\%,3)(P/F,12\%,2)](A/P,12\%,5)(.5) \\ &= -\$7{,}417 \end{aligned}$$

FMC:

$$\begin{aligned} EUCF &= -29{,}000(A/P,12\%,7) + 4{,}000(.5)(A/F,12\%,7) \\ &\quad - (.5)(4{,}000) + (A/P,12\%,7)(.5) \\ &\quad [29{,}000(.15)(P/F,12\%,1) + 29{,}000(.22)(P/F,12\%,2) \\ &\quad + 29{,}000(.21)(P/A,12\%,3)(P/F,12\%,2)] \\ &= -\$5{,}896 \end{aligned}$$

The FMC is more economical.

b)

That either truck can be (1) repeated identically into the indefinite future and (2) the service to be provided (material handling) is required forever.

10-9 **

A rent-a-car company, whose Minimum Attractive Rate of Return is 12% after taxes, is considering replacing part of its fleet. The cars in question were purchased 2 years ago (1979) for $8500 each. They have been depreciated under the tax laws prior to the Tax Act of 1981 using the Straight Line method, 5-year depreciable life and no salvage value for tax purposes. Each car has a remaining economic life of 2 years, no salvage value at that time, and the annual operation and maintenance expenses are estimated to be $7000.

New cars can be purchased from a local dealer at a cost of $10,000 each. Annual operation and maintenance costs would be $2000, an economic life of 4 years and no salvage value at that time. They would be depreciated using a recovery period of 3 years, and the Accelerated Cost Recovery System (ACRS) percentages: 25% the first year, 38% the second year and 37% the third year. For this company the effective income tax rate is 55% and the capital gains tax rate is 28%.

Write Equivalent Uniform Cash Flow (EUCF) equations from which you could compute the trade-in value the rent-a-car company would need on each old car to make the new cars economically attractive. Show the values of i and n in the equations, but do not waste time looking up the values of the factors.

****

The trade in value, TI, is that value such that $EUCF_{old} = EUCF_{new}$. The equations are:

$$EUCF_{old} = -7{,}000(.45)\,[1] + 1{,}700(.55)\,[2] + 1{,}700(.55)(A/F, 12\%, 2)\,[3]$$
$$- [.45(TI - 5{,}100) + TI](A/P, 12\%, 2)\,[4]$$

$$EUCF_{new} = -10{,}000(A/P, 12\%, 4)\,[5] - 2{,}000(.45)\,[6]$$
$$+ .55[10{,}000(.25)(P/F, 12\%, 1) + 10{,}000(.38)(P/F, 12\%, 2$$
$$+ 10{,}000(.37)(P/F, 12\%, 3)]\,[7]$$

1- A.T. Expenses

2- Depreciation tax savings. Depr = $8500/5 $1,700

3- Tax savings for 5th year depr. in year of disposal.

4- A.T. first cost of each old car. Book value of each old car at trade in would be 8,500 - 2(1,200) = $5,100.

5- First cost

6- A.T. Expense

7- Tax savings from depreciation

*** **10-10**

A graduate of an engineering economy course has compiled the following set of estimated costs and salvage values for a proposed machine with a first cost of $15,000; however, he has forgotten how to find the most economic life. Your task is to show him how to do this by calculating the Equivalent Uniform Annual Cost (EUAC) for n = 8, if the Minimum Attractive Rate of Return (MARR) is 15%. Ignore income taxes. You must show how to use gradients and uniform series factors in your solution.

Life (n) Years	Estimated End-of-Year Maintenance	Estimated Salvage if sold in Year n
1	$ 0*	$10,000
2	0*	9,000
3	300	8,000
4	300	7,000
5	800	6,000
6	1,300	5,000
7	1,800	4,000
8	2,300	3,000
9	2,800	3,000
10	3,300	3,000

*warranty period

Remember: Calculate only <u>one</u> EUAC (for n = 8). You are not expected to actually find the most economical life.

****

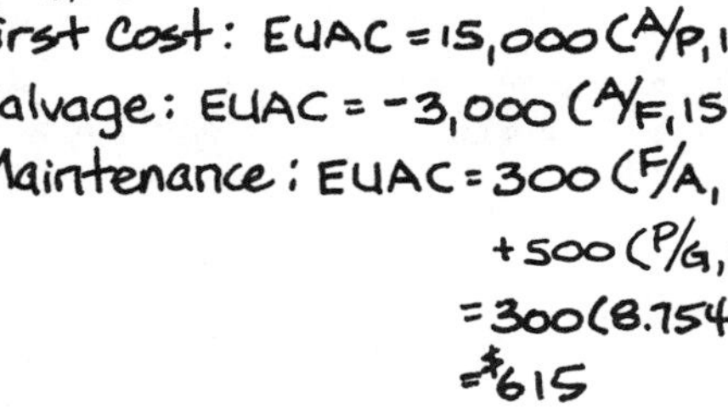

First Cost: EUAC = 15,000(A/P, 15%, 8) = 15,000(.2229) = $3,344

Salvage: EUAC = −3,000 (A/F, 15%, 8) = −3,000(.0729) = −$219

Maintenance: EUAC = 300 (F/A, 15%, 6)(A/F, 15%, 8)

+ 500 (P/G, 15%, 5)(P/F, 15%, 3)(A/P, 15%, 8)

= 300(8.754)(.0729) + 500(5.775)(.6575)(.2229)

= $615

Total $EUAC_8$ = $3,740

(A complete analysis would show that the most economic life is 7 years with $EUAC_7$ = 3,727)

10-11 **

The computer system used for production and administration control at a large cannery is being considered for replacement. Of the available replacement systems, "Challenger I" has been considered the best. However, it is anticipated that after one year, the "Challenger II" model will become available, with significant technological modifications. The salvage value projections for these three systems are summarized below. Assuming that their performance would otherwise be comparable, should we replace the existing system either this year or next year? Assume the MARR equals 12%, and a useful life of 5 years on all alternatives.

	Salvage Value at the end of the year		
Year	Existing Computer	Challenger I	Challenger II
0	$20,000	$25,000	$24,000
1	16,000	22,000	23,000
2	13,000	21,000	23,000
3	11,000	20,000	22,000
4	8,000	16,000	16,000
5	3,000	10,000	10,000

****

1. Existing Computer (defender)

YR.	Salv. Value at end of year	EUAC Capital Recovery $=(P-S)(A/P,12\%,n)+Si$
0	P=20k	–
1	16k	6,400
2	13K	5,701.9
3	11k	5,066.7
4	8K	4,910.4*
5	3k	5,075.8

* Economic Life 4yrs, EUAC=4,910.4

2. "Challenger I"

YR.	Salv. Value at End of Year	EUAC of Capital Recovery $=(P-S)(A/P,12\%,n)+Si$
0	P=25k	–
1	22K	6,000
2	21K	4,886.8
3	20k	4,481.5**
4	16K	4,882.8
5	10K	5,361

** Economic Life 3yrs, EUAC=4481.5

3. "Challenger II"

YR[(a)]	Salvage Value at end-of-year	EUAC Capital Recovery $= (P-S)(A/P,12\%,n)+Si$	
0	P = 24K	—	***
1	23K	3,880	Economic Life
2	23K	3,351.7****	2 yrs, EUAC = 3,351.7
3	22K	3,472.6	(a) year numbers do not refer to the same time scale as for Challenger I.
4	16K	4,553.6	
5	10K	5,083.6	

Note $EUAC_{chall\,II} < EUAC_{chall\,I} < EUAC_{defender}$, but, should we replace it now or should we wait one year for Challenger II?

Alternative A: Don't wait $\Rightarrow EUAC_A = EUAC_{chall\,I} = 4{,}481.5$

Alternative B: Wait one year before replacement.

$$EUAC_B = [\underbrace{6{,}400(P/A,12\%,1)}_{\text{cost of keeping defender one more year.}} + \underbrace{3{,}351.7(P/A,12\%,2)(P/F,12\%,1)}_{\text{cost of challenger II at its best (2 yrs. economic life)}}](A/P,12\%,3)$$

$$EUAC_B = [6{,}400(.8929) + 3{,}351.7(1.69)(0.8929)](.4163) = 4{,}484.49$$

Since $EUAC_A \approx EUAC_B$ we should preferably wait one year, although strictly speaking we can choose either option.

** 10-12**

An existing machine has operating costs of $300 per year and a salvage value of $100 (for all years). A new replacement machine would cost $1000 to purchase, and its operating cost over the next period of **t** years (not per year) is $M = 200t + 10t^2$. Assume i = zero percent.

(a) what is the most economic life t^* for the new machine?
(b) Should the old machine be replaced with the new one?

****

a) Cost per year $= AC = \frac{1{,}000}{t} + 200 + 10t \quad \left(= \frac{1000}{t} + \frac{M}{t}\right)$

$$\frac{dAC}{dt} = \frac{1{,}000}{t^2} + 10 = 0 \quad \Rightarrow t^* = \underline{+10 \text{ years}}$$

b) $AC^* = AC(10) = \frac{1{,}000}{10} + 200 + 10(10) = 400 \geq$ Annual cost of old machine for any number of years.

$\therefore$ No, keep old one.

10-13 **

A truck salesman is quoted as follows:

"Even though our list price has gone up to $42,000, I'll sell you a new truck for the old price of only $40,000, an immediate savings of $2,000, and give you a trade-in allowance of $21,000, so your cost is only ($40,000 - 21,000) = $19,000. The book value of your old truck is $12,000, so you're making an additional ($21,000 - 12,000) = $9,000 on the deal." The salesman adds, "Actually I am giving you more trade-in for your old truck than the current market value of $19,500, so you are saving an extra ($21,000 - 19,500) = $1,500."

(a) In a proper before-tax replacement analysis, what is the first cost of the defender?

(b) In a proper before-tax replacement analysis, what is the first cost of the challenger?

(c) What is the proper first cost to use for the defender in an after-tax analysis? The corporation's incremental tax rate is 30%.

(d) What is the proper first cost to use for the challenger in an after-tax analysis if no investment tax credit is taken?

****

a) $19,500. The defender 1st cost before taxes is always the current market value, not trade-in or book value.

b) $38,500. With an inflated trade-in of $1,500 (21,000 - 19,500) the new truck can be purchased for $40,000. Therefore, the appropriate value for a replacement analysis is:
$40,000 - $1,500 = $38,500

c) $17,250. The after-tax 1st cost is the current after-tax value if it were sold now. The taxes are affected by an ordinary gain of 19,500 (CMV) - 12,000 (BV). Thus

BTCF	Tax Inc.	Taxes	ATCF
19,500	7,500	2,250	17,250

d) $38,500. Without an investment tax credit the after-tax 1st cost is the same as the before-tax value. There is no tax impact.

11 INFLATION

** **11-1**

A European investor lives near to one of his country's borders. In Country A (where he lives), an 8% interest rate is offered in banks, and the inflation rate is 3%. Country B, on the other hand, has an inflation rate of 23%, and their banks are offering 26% interest on deposits.

(a) What is the real or effective interest rate that this person gets when investing in his Country A?

(b) This investor believes that the currency of Country B will not change in its value relative to the value of the currency of Country A during this year. In which country would he get a larger effective interest rate?

(c) Suppose now that he invests in a bank in Country B, and that his prediction was wrong. The currency of Country B was devaluated 20% with respect to the exchange value of Country A's currency. What is the effective interest rate that he obtained?

****

a) $i' = ?$ if $i_A = 8\%$, $f_A = 3\%$

$i = i' + f + i'f$

$.08 = i' + .03 + i'(.03) \Rightarrow i' = \frac{0.05}{1+.03} = 0.0485 \therefore 4.85\%$

b) if investing in Country A: $i'_A = 0.0485$

if investing in Country B: $i_B = 26\%$, $f_A = 3\%$ (note that he lives in Country A. Inflation of Country B does not affect him directly)

$i'_B = \frac{i_B - f_A}{1+f_A} = \frac{0.26 - 0.03}{1+0.03} = 0.2233 \therefore 22.33\%$

He can get a larger effective interest rate in Country B. (A person that invests and lives in Country B, however, gets $i' = \frac{i_B - f_B}{1+f_B} = \frac{0.26-0.23}{1+0.23} = 0.02439$, much less than a person who lives and invests in A).

c) x = amt. originally invested in B (measured in curreny A).
The amt. collected at end of 1 yr (measured in currency A):
$(1.0-0.2)(1.26x) = 1.008x$
(1.0−0.2): due to the devaluation; (1.26x): due to initial deposit(+) interest

the interest is then $i = \frac{1.008x - x}{x} = 0.008$
but during that year the inflation in Country A (where he lives) was 3%, therefore
$i = 0.008$
$f = 0.03$ $\qquad i' = \frac{0.008-0.03}{1+.03} = -0.02136$
$i' = ?$

He actually lost money (negative effective interest rate of −2.136%)

11-2 **

The first sewage treatment plant for Athens, Georgia cost about $2 million in 1964. The utilized capacity of the plant was 5 million gallons/day (mgd). Using the commonly accepted value of 135 gallons/person/day of sewage flow, find the cost per person for the plant. Adjust the cost to 1984 dollars with inflation at 6%. What is the annual capital expense per person if the useful life is 30 years and the value of money is 10%?

****

Population equivalents $= \frac{5 \text{ mgd}}{135} = 37{,}037$

Cost per capita $= \frac{\$2{,}000{,}000}{37{,}037} = \$54.$

1984 dollars, $F = 54(F/P, 6\%, 20)$
$= 54(3.207) = \$173.18$

Annual Cost, $A = 173.18(A/P, 10\%, 30)$
$= 173.18\,(.1061) = \$\underline{18.37}$

** **11-3**

How much life insurance should a person buy if he wants to leave enough money to his family, so they get $25,000 per year in interest, of constant Year 0 value dollars? The interest rate expected from banks is 11%, while the inflation rate is expected to be 4% per year.

****

The actual (effective) rate that the family will be getting is $i' = \frac{i-f}{1+f} = \frac{0.11-0.04}{1.04} = 0.0673 \therefore 6.73\%$

To calculate P, $n=\infty$ (capitalized cost)

$P = \frac{A}{i'} = \frac{25,000}{0.0673} = 371,471$

Therefore, he needs to buy about 370,000 of life insurance

** **11-4**

Property, in the form of unimproved land, is purchased at a cost of $8000 and is held for six years when it is sold for $32,600. An average of $220 each year is paid in property tax and may be treated at an interest of 12%. The long-term capital gain tax is 15% of the long-term capital gain. Inflation during the period is treated as 7% per year. What is the annual rate of return for this investment?

****

Long term gains = 32,600 − 8,000 = 24,600

Tax on Long term gains = .15 × 24,600 = 3,690.

Property tax = 220(F/A, 12%, 6) = 220(8.115) = 1,785.30

Adjusted FW = 32,600 − 3,690 − 1,785.30 = 27,624.70

also $FW = 8{,}000(1+i_{eq})^6$

$\therefore (1+i_{eq}) = \left(\frac{27,124.70}{8,000}\right)^{1/6} = 1.2257$

$(1+i_{eq}) = (1+i)(1+i_f)$, note: inflation and interest rates are additive in effect.

$1+i = \frac{1.2257}{1.07} = 1.1455$ or <u>14.6% rate of return</u>

11-5 **

The auto of your dreams costs $20,000 today. You have found a way to earn 15% tax free on an "auto purchase account." If you expect the cost of your dream auto to increase by 10% per year, how much would you need to deposit in the "auto purchase account" to provide for the purchase of the auto 5 years from now?

****

Cost of Auto 5 years hence (F) = $P(1 + \text{inflation rate})^n$

$= 20{,}000(1+0.10)^5 = 32{,}210$

Amount to deposit now to have $32,210 five years hence.

$P = F(P/F, i\%, n) = 32{,}210(P/F, 15\%, 5)$

$= 32{,}210(0.4972) = \underline{16{,}014.81}$

11-6 **

On January 1, 1975 the National Price Index was 208.5, and on January 1, 1985 it was 516.71. What was the inflation rate, compounded annually, over that 10-year period? If that rate contines to hold for the next 10 years, what National Price Index can be expected on January 1, 1995?

****

NPW = 0

$0 = -208.5 + 516.71(P/F, \text{inflation rate}, 10)$

$(P/F, \text{inflation rate}, 10) = \frac{208.5}{516.71} = 0.4035$

Trial & Error Solution:

Try $i = 9\%$: $-208.5 + 516.71(0.4224) = +9.76$

Try $i = 10\%$: $-208.5 + 516.71(0.3855) = -9.31$

Interpolation: Inflation Rate $= 9\% + (1\%)\left(\frac{9.76}{9.76 - (-9.31)}\right)$

$= 9.51\%$

National Price Index$_{1995}$ $= 516.71(1+0.0951)^{10} = \underline{1{,}281.69}$

** **11-7**

A department store offers two options to buy a new color TV which has a price of $440.00. A customer can either pay cash and receive immediately a discount of $49.00 or he can pay for the TV on the installment plan. The installment plan has a nominal interest rate of 12% compounded bi-yearly and would require an initial down payment of $44.00 followed by four equal payments (principal and interest) every six months for two years.

If for the typical customer the real Minimum Attractive Rate of Return is 5%, what is the maximum effective annual inflation rate for the next two years that would make paying cash preferred to paying installments? All figures above are quoted in time zero dollars.

****

The monthly payments in nominal dollars if the installment plan was selected would be

$$(-\$440 + 44)(A/P, \tfrac{.12}{2}, 4) = -\$114.28$$

The breakeven inflation rate is that rate such that $NPV_{Buy} = NPV_{Install}$. or $NPV_{Buy-Install} = 0$

$$NPV_{B-I} = ((-440 + 49) + 44) + 114.28(P/A, i_{1/2}, 4) = 0$$

$(P/A, i_{1/2}, 4) = 3.0364$ therefore the nominal effective semi-annual MARR would have to be $i_{1/2} \approx .115$. The nominal effective annual rate would be $i = (1.115)^2 - 1 = 0.2432$. The effective annual inflation rate can now be computed from the formula $(1.2432) = (1.05)(1+f)$; thus $f = .1840$

** **11-8**

An automobile that cost $6500 in 1980 has an equivalent model four years later in 1984 that cost $9250. If inflation is considered the cause of the increase, what was the average annual rate of inflation?

****

$$F = P(1+i_f)^n$$

$$9{,}250/6{,}500 = (1+i_f)^4$$

$$1 + i_f = (1.423)^{1/4} = 1.092$$

$$\underline{i_f = 9.2\%}$$

11-9 **

A machine has a first cost of $100,000 (in today's dollars) and a salvage value of $20,000 (in then-current dollars) at the end of its ten year life. Each year it will eliminate one full-time worker. A worker costs $30,000 (today's dollars) in salary and benefits. Labor costs are expected to escalate at 10% per year. Operating and maintenance costs will be $10,000 per year (today's dollars) and will escalate at 7% per year.

Construct a table showing before-tax cash flows in current dollars, and in today's dollars. The inflation rate is 7%.

****

End of	Current Dollars				Today's
Year	Savings	O&M	Capital	Total	Dollars
0			-100,000	-100,000	-100,000
1	33,000	-10,700		22,300	20,841
2	36,300	-11,449		24,851	21,706
3	39,930	-12,250		27,680	22,595
4	43,923	-13,108		30,815	23,509
5	48,315	-14,026		34,290	24,448
6	53,147	-15,007		38,140	25,414
7	58,462	-16,058		42,404	26,407
8	64,308	-17,182		47,126	27,428
9	70,738	-18,385		52,354	28,477
10	77,812	-19,672	20,000	78,141	39,123

11-10 **

A project has been analyzed assuming 6% inflation and is found to have a monetary Internal Rate of Return (IRR) of 22%. What is the real IRR for the project?

****

Real IRR = (1.22)/(1.06) - 1 = 0.1509 or 15.09%

** **11-11**

A solar energy book gives values for a solar system as follows: Initial cost, $6500; Initial fuel savings, $500/year; Expected life, 15 years; Value of money, 10%; Inflation, 12%; and Incremental income tax rate, 25%. If we define the payback condition as the time required for the present worth of the accumulated benefit to equal the accumulated present worth of the system cost, what is the time required to reach the payback condition? Since the income tax benefit is related to the annual interest expense, treat it as a reduction of the annual cost.

****

Annualizing P: $A = 6{,}500\,(A/P, 10\%, 15)$
$= 6{,}500\,(.1315) = 854.75$

$1 + i_c = (1.10)(1 + 0.25 \times 0.10) = 1.1275$

PW of costs $= 854.75(P/A, 12.75\%, 15)$
$= 854.75(6.547) = 5{,}595.82$

$$1 + i_{eq} = \frac{1 + i_f}{1 + i} = \frac{1.12}{1.10} = 1.018$$

The solution strategy is to find the time for the PW of benefits to equal PW of cost. When the combined effect of the two rates on a distributed A amount are opposed then the net effect retains the direction of the larger rate. The inflation rate is greater than the time value of money, which is abnormal. To solve this problem, find the PW of benefit, and to do that we must get FW of the equivalent rate, i_{eq}.

Try 10 years: FW $= 500(F/A, 1.8\%, n)$
$= 500(10.850) = 5{,}425.06$

Try 11 years: FW $= 500(F/A, 1.8\%, n)$
$= 500(12.045) = 6{,}022.72$

Interpolate payback = <u>10.3 years</u>

** **11-12**

A company requires a real MARR of 12%. What monetary MARR should they use if inflation is expected to be 7%?

****

Monetary MARR $= (1.12)(1.07) - 1 = 0.1984$ or 19.84%

11-13 ***

An asset with a three year UOP* life will be purchased for \$20,000. It will result in savings (net annual benefits) of \$9600, \$11,520, \$13,824, and \$16,589 in years 1 through 4, respectively, after which it will be sold for \$3000. All amounts are in actual dollars. The company's incremental tax rate is 50%. Ignore the investment tax credit. The company uses a real Minimum Attractive Rate of Return (MARR) of 20% and a monetary MARR of 32%. Recaptured depreciation (gain on disposal) is taxed as ordinary income. What is the after-tax present worth?

* UOP Depreciation: 29%, 47%, and 24% for years 1, 2, and 3.

****

EOY	ACTUAL DOLLAR BTCF	3-YR UOP Depr	TAXABLE INCOME	CASH FLOW FOR TAXES	ACTUAL DOLLAR ATCF	YEAR 0 DOLLARS ATCF
0	-20,000				-20,000	-20,000
1	9,600	5,800	3,800	-1,900	7,700	7,000
2	11,520	9,400	2,120	-1,060	10,460	8,644.63
3	13,824	4,800	9,024	-4,512	9,312	6,996.24
4	16,589		16,589	-8,294.5	8,294.5	5,665.26
	3,000	(0*)	3,000**	-1,500	1,500	1,024.52

$PW(32\%) = -20{,}000 + 7{,}700(1.32)^{-1} + \cdots + (8{,}294.5 + 1{,}500)(1.32)^{-4}$
$= -888.54$

$PW(20\%) = -20{,}000 + 7{,}000(1.20)^{-1} + \cdots + (5{,}665.26 + 1{,}024.52)(1.20)^{-4}$
$= -888.53$

The conversion from Actual Dollar ATCF to Year 0 Dollar ATCF was made by noting that inflation rate, f is given by: $f = 1.32/1.20 - 1 = 0.1$ or 10%

*Book value

**Ordinary income. See the note at the end of the Income Tax chapter.

11-14 ***

The real interest rate is 4%.
The inflation rate is 8%.
What is the apparent interest rate?

****

$i = i' + f + i'f$
$= 0.04 + 0.08 + 0.04 \times 0.08 = 12.32\%$

** **11-15**

Compute the internal rate of return based in constant (Year 0) dollars for the following after-tax cash flow given in current or actual dollars. Inflation is assumed to be 7% per year. (Round to the nearest dollar.

Year	After Tax Cash Flow in actual dollars
1988 (Yr 0)	-$10,000
1989	+3,745
1990	+4,007
1991	+4,288
1992	+4,588

****

Year	After Tax Cash Flow in Constant Dollars
1988 (0)	-10,000
1989 (1)	$3{,}745/1.07 = 3{,}500$
1990 (2)	$4{,}007/(1.07)^2 = 3{,}500$
1991 (3)	$4{,}288/(1.07)^3 = 3{,}500$
1992 (4)	$4{,}588/(1.07)^4 = 3{,}500$

$$NPW = 0 = -10{,}000 + 3{,}500(P/A, i\%, 4)$$
$$(P/A, i\%, 4) = 10{,}000/3{,}500 = 2.857$$
$$ROR = i = 15\%$$

** **11-16**

The capital cost of a wastewater treatment plant for a small town of about 6000 people was estimated to be about $85/person in 1969. If a modest estimate of the rate of inflation is 5.5% for the period to 1984, what is the per capita capital cost of a treatment plant now?

****

$$F = P(1+i_f)^n$$
$$= 85(1+.055)^{15}$$
$$= 85(2.232) = \$189.76$$

11-17 **

A lot purchased for $4500 is held for five years and sold for $13,500. The average annual property tax is $45 and may be accounted for at an interest rate of 12%. The income tax on the long term capital gain is at the rate of 15% of the gain. What is the rate of return on the investment if the allowance for inflation is treated at an average annual rate of 7%?

****

Long term gain = 13,500 - 4,500 = 9,000

Tax on long term gain = (.15)(9,000) = 1,350

Property tax = 45(F/A, 12%, 5)

= 45(6.353) = 285.89

Adjusted return = 13,500 - 1,350 - 285.89 = 11,864.12

also = $4{,}500(1+i_{eq})^5$

$$1+i_{eq} = \left(\frac{11{,}864.12}{4{,}500}\right)^{1/5} = 1.214$$

$$1+i = \frac{1+i_{eq}}{1+i_f} = \frac{1.214}{1.075} = 1.129 \; ; \; \underline{i = 12.9\%}$$

11-18 **

Undeveloped property near the planned site of an interstate highway is estimated to be worth $48,000 in six years when the construction of the highway is completed. Consider a 15% capital gains tax on the gain, an annual property tax of 0.85% of the purchase price, an annual inflation rate of 7%, and an expected return of 15% on the investment. What is the indicated maximum purchase price now?

****

Let x = purchase cost

$1+i_{eq} = (1.15)(1.07) = 1.231$

Annual property tax = $.0088x$

FW of property tax = $.0085x$ [F/A, 23.1%, 6]

= $.0909x$

Adj. return = $48{,}000 - .15(48{,}000 - x) - .0909x$

also = $x(1.231)^6 = 3.48x$

$40{,}800 + .15x - .0909x = 3.48x$

$\underline{x = \$11{,}927 \text{ purchase cost}}$

** **11-19**

Minor Oil Co. owns several gas wells and is negotiating a 10-year contract to sell the gas from these wells to Major Oil Co. They are negotiating on the price of the gas the first year, $ per thousand cubic feet (MCF), and on the escalation clause, the percentage rate of increase in the price every year thereafter. Minor expects the wells to produce 33,000 MCF the first year and to decline at the rate of 18% every year thereafter. Minor has agreed to spend $500,000 now to lay pipelines from each well to Major's nearby refinery. What should the minimum price be the first year and what should the escalation rate be if Minor wants their revenue each year to remain constant (uniform) over the life of the contract. Assume an end-of-year convention and a Minimum Attractive Rate of Return (MARR) of 15%.

****

Required income to earn the 15% MARR on $500,000:

EUAC = 500,000 (A/P, 15%, 10) = 500,000(0.1993) = 99,650

Initial price = $99,650/33,000 MCF = $3.02/MCF

Annual production declines to (1−0.18) of initial rate each year.

Let f = required annual escalation rate

Then $(1-0.18)(1+f)=1$ to keep the revenue constant

$$f = \frac{1}{(1-0.18)} - 1 = 0.2195/\text{year}$$

** **11-20**

Jack purchases a lot for $40,000 cash and plans to sell it after 5 years. What should he sell it for if he wants a 20% before-tax rate of return, after taking the 5% annual inflation rate into account?

****

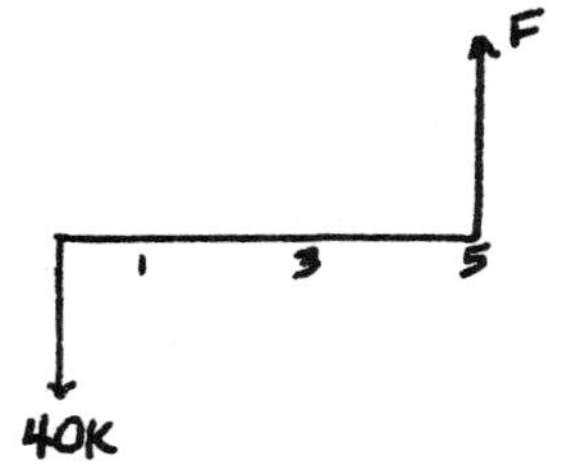

F = 40K (F/P, 20%, 5)(F/P, 5%, 5)

= 40K (2.488)(1.276) = <u>126,988</u>

11-21 **

A solar system costs $6500 initially and qualifies for a federal tax credit (40% of cost, not to exceed $4000). The cost of money is 10%, and inflation is expected to be 7% during the life of the system. The expected life of the system is 15 years with zero salvage value. The homeowner is in the 40% income tax bracket. The initial fuel saving is estimated at $500 for the first year and will increase in response to inflation. The annual maintenance cost of the system is estimated at 5% of the annualized cost of the system. What is the time required for the payback condition to be reached for this investment?

****

Adjust initial cost by tax credit: $P = .60(6{,}500) = 3{,}900$

Annualize cost: $A = 3{,}900\,(A/P, 10\%, 15)$
$= 3{,}900\,(.1315) = 512.85$

$1 + i_c = \dfrac{1.10(1 + .40 \times .10)}{1.05} = 1.0895$ { $1 + i_m = 1.05$ represents maintenance charge as a rate

PW of costs $= 512.85\,(P/A, 8.95\%, 15)$
$= 512.85\,(8.086) = 4{,}146.67$

$1 + i_{eq} = 1.10/1.07 = 1.028$

Try 9 years: $PW = 500\,(P/A, 2.8\%, n)$
$= 500\,(7.868) = 3{,}934.18$

Try 10 years: $PW = 500\,(P/A, 2.8\%, n)$
$= 500\,(8.618) = 4{,}308.97$

Interpolate payback = <u>9.6 years</u>

11-22 **

Acme Company is considering the purchase of a new machine for $30,000 with an expected life of 20 years when it is estimated the salvage value will be zero. An investment tax credit of 7% will be allowed. The incremental tax rate (tax bracket) for the company is 36%, inflation is estimated to be 7%, and the value of money is 12%. If the annual benefit is estimated to be $2500 per year over current production and maintenance costs, what will be the time required for the payback condition to be reached (that is, the point where the present worth of benefits up to payback equals the present worth of costs for the life expectancy of the equipment)?

****

Tax credit reduces initial cost $P = 30{,}000 \times .93 = 27{,}900$

Annualize P: $A = 27{,}900(A/P, 12\%, 20)$

$= 27{,}900(.1339) = \$3735.81$

$1 + i_c = (1+i)(1+.36 \times .12) = 1.1684$

i_c combines: interest i with income tax credit rate

PW of cost $= 3{,}735.81(P/A, 16.84\%, 20)$

$= 3{,}735.81(5.674^*) = 21{,}197$

$1 + i_{eq} = (1+i)/(1+i_f)$

$= 1.12/1.07 = 1.0467$

Try 11 years: $PW = 2{,}500(P/A, 4.67\%, n) = 21{,}130^*$

Try 12 years: $PW = 2{,}500(P/A, 4.67\%, n) = 22{,}575^*$

Interpolate payback = <u>11.05 years</u>

* numbers found by interpolation

** **11-23**

The net cost of a solar system for a home is $8000 and it is expected to last 20 years. If the value of money is 10%, inflation is expected to be 8%, and the initial annual fuel saving is $750, what is the time for the payback condition to be reached for the system? Assume the homeowner is in the 30% income tax bracket.

****

Annualize P: $A = 8{,}000(A/P, 10\%, 20)$

$= 8{,}000(.1175) = 940$

$1 + i_c = (1.10)(1 + .10 \times .30) = 1.133$

PW of Cost $= 940(P/A, 13.3\%, 20)$

$= 940(6.900) = 6{,}486$

$1 + i_{eq} = 1.10/1.08 = 1.0185$

Try 9 years: $PW = 940(P/A, 1.85\%, n)$

$= 940(8.228) = 6{,}171$

Try 10 years: $PW = 940(P/A, 1.85\%, n)$

$= 940(9.053) = 6{,}790$

Interpolate payback = <u>9.5 years</u>

11-24 **

An undeveloped parcel of land in Clarke County, Georgia was purchased in 1980 for $4850. The property tax was $8 for the first year and is assumed to have increased by $2 per year. The capital gain tax is 13.6% of the long term capital gain. Inflation for the period is treated at an 8% annual rate. A 16% rate of return on the investment is desired. What is the indicated sale price in 1985?

****

$1 + i_{eq} = (1.16)(1.08) = 1.2528$

FW of property tax $= [8 + 2(A/G, 25.28\%, 5)][F/A, 25.28\%, 5]$

$= [8 + 2(3.12)][8.252] = 91.74$

Let x = selling price

Long term capital gains tax $= 0.136(x - 4{,}850) =$

$= .136x - 659.60$

Adj. ret $= x - [.136x - 659.60 + 91.74]$

$= .864x + 567.86$

also $= 4{,}850(1.2528)^5 = 14{,}967.54$

$.864x = 14{,}399.68$

$x =$ $16,666.31 selling price

11-25 **

A company has designed a VLSI circuit and a production system to manufacture it. It is believed that it can sell 100,000 circuits per year if the price in then-current dollars is cut 20% per year (for example, if the unit price in the first year is $100, then the price in years 2 through 5 would be $80, $64, $51.20, and $40.96).

The required revenue for the five years is $2,500,000 per year in today's dollars. The real and monetary costs of capital are 8.8% and 16.416%, respectively. What should the then-current dollar selling price be in each of the years 1 through 5?

****

Let R_1 be the required revenue in year 1, then the required revenue in years 2 through 5 is $.8R_1$, $.64R_1$, $.512R_1$, and $.4096R_1$. Since these are in then-current \$,

$$(2{,}500{,}000)(P/A, 8.8\%, 5) = R_1(1.16416)^{-1} + 0.8R_1(1.16416)^{-2} + 0.64R_1(1.16416)^{-3} + 0.512R_1(1.16416)^{-4} + 0.4096R_1(1.16416)^{-5}$$

$9{,}774{,}800 = 2.32523R_1$

$R_1 = 4{,}203{,}804$ or a unit price of \$42.04 in yr. 1
\$33.63 in yr. 2
\$26.90 in yr. 3
\$21.52 in yr. 4
\$17.22 in yr. 5

** **11-26**

An investment in undeveloped land of \$9000 was held for four years and sold for \$21,250. During this time property tax was paid that was, on the average, 0.4% of the purchase price. Inflation in this time period averaged 7% and the income tax was 15.2% of the long term capital gain. What rate of return was obtained on the investment?

****

Long term capital gains tax $= 0.152(21{,}250 - 9{,}000) = \$1{,}862$

Prop. tax $= .004 \times 9{,}000 = \$36/\text{year}$

FW of prop. tax $= 36(F/A, i_{eq}, 4)$

$$1 + i_{eq} = \left(\frac{21{,}250 - 1{,}862}{9{,}000}\right)^{1/4} = 1.2115 \text{ (1st estimate)}$$

FW of prop. tax $= 36(F/A, 21.15\%, 4) = 36(5.47) = \$197.$

$$1 + i_{eq} = \left(\frac{21{,}250 - 1{,}862 - 197}{9{,}000}\right)^{1/4} = 1.2084 \text{ (2nd estimate)}$$

$$\text{Rate of Return} = \frac{1.2084}{1.07} - 1 = \underline{12.9\%}$$

check: $9{,}000(1.2084)^4 = \$19{,}191$
prop. tax = 197
L.T. cap. gain tax = 1,862
\$21,250

11-27 **

Steve Luckee has just been informed that he has won $1 million in a local state lottery; unfortunately the Internal Revenue Service has also been so informed. Due to tax considerations, Steve has a choice in the manner in which he will receive his winnings. He can either receive the entire amount today, in which case his combined federal and state income tax rate will soar to 74%, or he can receive the money in uniform annual payments of $50,000 over the next 20 years, beginning one year from today, in which case his annual payments will be taxed at a lower rate. Steve expects inflation to be a problem for the next 20 years which will obviously reduce the value of his winnings if takes the annual payment option. Steve uses a real dollar MARR of 5% in his decisions.

(a) If Steve estimates that the inflation rate **f** would be 6% and his combined federal and state tax rate **d** would be 40% for the next 20 years, what action should he take?

(b) What would be the present worth of the cost of an error in Steve's estimates in (a) if the true values eventually prove to be **f** = 8% and **d** = 20%?

(c) If Steve is certain that 6% $\le$ **f** $\le$ 8% and 20% $\le$ **d** $\le$ 40% for the next 20 years, what is the maximum cost of errors in estimating **f** and **d**?

****

a) The value of the entire amount is:

(1-.74)($1M) = R$260,000.

The real NPV of the payments is:

NPV = (1-.4)($1M/20)(P/A, (1.05)(1.06)-1, 20)

= R$234,288

Take the entire amount today, it is worth more.

b) The real NPV of the payments would be:

NPV = (1-.2)(50k)(P/A, (1.05)(1.08)-1, 20)

= R$274,370.

which is more than the entire amount today. Steve's errors in estimation would cost him 274,370 - 260,000 = R$14,370.

c) The maximum cost of an error will occur one of two ways, either Steve's estimates indicates he shou[ld] take the entire amount today when he shouldn't or that he take the payments when he shouldn't. If he takes the entire amount when he shouldn't, then for f = .06 and d = .2, and the payments are worth NPV = .8(50K)(P/A, (1.05)(1.06)-1, 20) = R$312,384 and the cost is \$52,384. If he takes the payments when he shouldn't, then f = .08 and d = .4, and the payments are worth .6(50K)(P/A, (1.05)(1.08)−1, 20) = R\$205,777, and the cost is \$54,223. The maximum cost is \$54,223.

** **11-28**

An investor is considering the purchase of a bond. The bond has a face value of \$1000, a coupon rate (interest rate) of i = 0.06 compounded annually, pays interest once a year, and matures in 8 years. This investor's real MARR is 25%.

(a) If the investor expects an inflation rate of 16% per year for the next 8 years, how much should he be willing to pay for the bond?

(b) If, however, the investor expects a deflation rate of 18% per year for the next 8 years, how much should he be willing to pay for the bond?

****

a)

The net cash flows for bond in nominal dollars are

t	Net cash flows
0	N\$F.C.
1-8	+60 = .06(1,000)
8	+1,000

If the investor wants to earn a real 25%, but expects inflation at 16%, his nominal MARR must be (1.25)(1.16) − 1 = .45, and therefore

NPV = 0 = F.C. + 60(P/A, .45, 8) + 1,000(P/F, .45, 8)

F.C. = −177

He should be willing to pay no more than \$177.

b)

If the deflation rate (a negative inflation rate) is 18% then the real MARR is (1.25)(0.82)−1 = .025, and F.C. = −60(P/A, .025, 8) − 1,000(P/F, .025, 8) = −\$1,251.

11-29 **

An asset with a three year ACRS life (use ACRS percentages of 25%, 38%, and 37%, respectively) will be purchased in 198X for $10,000 (198X dollars). It will result in savings (net annual benefits) of $4000 per year (198X dollars) for three years, after which it will have a salvage value of $1000 (198X dollars).

Calculate the year by year after tax cash flows in 198X dollars if the company's incremental tax rate is 50% and the inflation rate is 10%. Assume that benefits inflate at the general inflation rate. Ignore the investment tax credit. Any gain on disposal (to the extent of the depreciation) is treated as ordinary income.

****

End of Year	0	1	2	3	
198X Dollar Before Tax Cash Flow	-10,000	4,000	4,000	4,000	1,000
Actual Dollar Before Tax Cash Flow	-10,000	4,400	4,840	5,320	1,331
ACRS Depreciation		2,500	3,800	3,700	(0*)
Taxable Income		1,900	1,040	1,624	1,331**
Cash Flow For Taxes		-950	-520	-812	-665.50
Actual Dollar After Tax Cash Flow	-10,000	3,450	4,320	4,512	665.50
198X Dollars After Tax Cash Flow	-10,000	3,136.36	3,570.25	3,389.93	500

The internal rate of return of the ATCF in actual dollars (A$) is 13.135%. This is the monetary rate of return. The internal rate of return of the ATCF in 198X dollars (R$) is 2.85%. This is the real rate of return. They are related by (1.13135) = (1.1)(1.0285)

The entry (0) in the ACRS Depreciation row is a reminder that the book value is zero at the end of three years.

**Ordinary income. See the note at the end of the Income Tax chapter.

12 BENEFIT-COST RATIO

** 12-1

A city engineer is considering installing an irrigation system. He is trying to decide which one of two alternatives to select. The two alternatives have the following cash flows:

Year	A	B
0	-$15,000	-$25,000
1-10	+5,310	+7,900

If interest is 12%, which alternative should the engineer select? Assume no salvage value. Use Incremental benefit/cost analysis.

****

	A	B
PW_C	-15,000	-25,000
PW_B	5,310 (P/A, 12%, 10) = 5310 × 5.65 = $30,000	7,900 (P/A, 12%, 10) = 7,900 × 5.65 = 44,635
B/C	30,000/15,000 = 2	44,635/25,000 = 1.79

B − A : PW_C = -10,000

Unif. Ann. Ben = +2,590

PW_B = 2590 × 5.65 = 14,634

$$\frac{\Delta B}{\Delta C} = \frac{14,634}{10,000} = 1.46 > 1$$

∴ choose B, higher cost alternative

12-2 **

With interest at 10%, what is the benefit-cost ratio for this government project?

Initial cost	$200,000
Additional costs at end of years 1 & 2	30,000/yr
Benefits at end of years 1 & 2	0
Annual benefits at end of years 3-10	90,000/yr

****

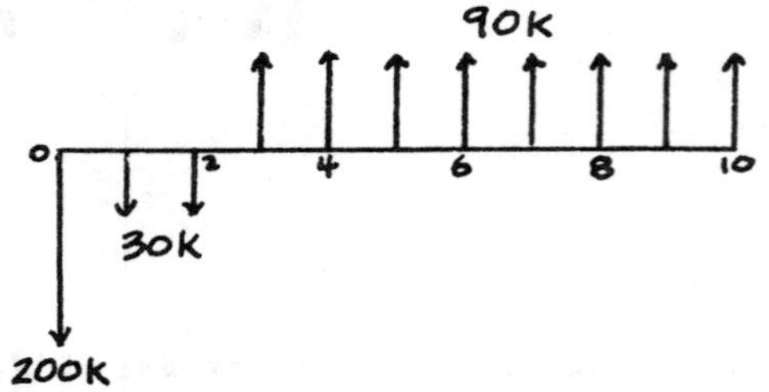

PW Cost = 200K + 30K (P/A, 10%, 2)

= 200K + 30K (1.736) = 252,080.

PW Benefits = 90K (P/A, 10%, 8)(P/F, 10%, 2)

= 90K (5.335)(.8264) = 396,800.

B/C = 396,800/252,080 = 1.574

12-3 **

A van is bought for $15,000 in Year 0. It generates revenues of $25,000 per year and expenses of $20,000 per year in each of the Years 1 through 5. Its salvage value is zero after 5 years. Ignoring income taxes, compute the Benefit/Cost ratio if **i** = 10%.

****

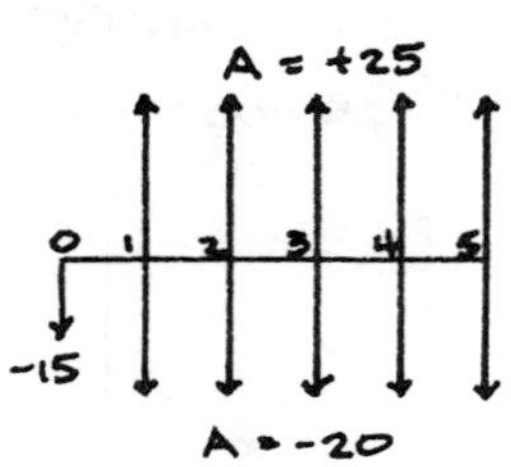

PWB = 25,000 (P/A, 10%, 5) = 25,000 (3.791)

= $94,775.

PWC = 20,000 (P/A, 10%, 5) + 15,000

= 20,000 (3.791) + 15,000 = $90,820

$$\frac{PWB}{PWC} = \frac{94,775}{90,820} = \underline{1.0435}$$

** **12-4**

A tax exempt municipality is considering the construction of an impoundment for city water supplies. Two different sites have been selected as technically, politically, socially, and financially feasible. The city council has asked you to do a benefit-cost analysis of the alternatives and recommend the best site. The city uses a 6% interest rate in all analyses of this type.

Year	Rattlesnake Canyon	Blue Basin
0	-$15,000,000	-$27,000,000
1-75	+2,000,000/yr	+3,000,000/yr

Which should you recommend?

****

		Rattlesnake	Blue Basin
PW of Ben / PW of Cost	=	$\frac{2\times10^6(P/A,6\%,75)}{15\times10^6}$	$\frac{3\times10^6(P/A,6\%,75)}{27\times10^6}$
B/C	=	2.19 > 1 ok	1.83 > 1 ok

YR	BB−RC
0	−12,000,000
1-75	+1,000,000

$$\frac{\Delta B}{\Delta C} = \frac{1\times10^6(P/A,6\%,75)}{12} = 1.37 > 1$$

∴ choose higher cost alternative, Blue Basin

** **12-5**

Two economists are arguing about the best way to rank projects under a capital constraint. One thinks the best way to do it is by NPW/PW of Cost. The other thinks the best way is to use B/C. Prove that both are essentially using the same technique.

****

NPW = PW of Benefits − PW of Cost

$$\frac{NPW}{PW\text{ of Cost}} = \frac{PW\text{ of }B - PW\text{ of }C}{PW\text{ of }C} = \frac{PW\text{ of }B}{PW\text{ of }C} - 1$$

$$\frac{B}{C} = \frac{PW\text{ of }B}{PW\text{ of }C}$$

$$\frac{NPW}{PW\text{ of }C} = \frac{B}{C} - 1$$

note: the constant −1 has no effect on the ranking

12-6 **

A consulting engineer is deciding which of two computers to purchase. Using Benefit-Cost analysis, which alternative should be selected if the interest rate is 10% per year?

Computer	Cost	Annual Benefit	Salvage	Useful Life
A	$48,000	$13,000	$0	6 yrs
B	40,000	12,000	0	6

****

Alternative A:

PW(B) = 13,000(P/A,10%,6) = 13,000(4.355) = $56,615.

PW(C) = 48,000

$B/C = 56,615/48,000 = 1.179$

Alternative B:

PW(B) = 12,000(P/A,10%,6) = 12,000(4.355) = $52,260

PW(C) = 40,000

$B/C = 52,260/40,000 = 1.3065$

Select B largest B/C ratio

$$\frac{\Delta B}{\Delta C} = \frac{56,615 - 52,260}{48,000 - 40,000} = \frac{4,355}{8,000} = .544$$

$\frac{\Delta B}{\Delta C} < 1.0$ select smallest cost alternative B

NPW = PW(B) − PW(C)

Da = 56,615 − 48,000 = +$8,615

Db = 52,260 − 40,000 = +$12,260

To maximize NPW − choose largest positive number alternative B

12-7 **

Given four mutually exclusive alternatives, each with a useful life of 20 years and no salvage value. Which alternative should be selected?

	A	B	C	D
PW of Costs	$4,000	$ 9,000	$6,000	$2,000
PW of Benefits	6,480	11,250	5,700	4,700

****

	A	B	C	D
B/C	1.62	2.35	0.95	2.35

exclude "C"

(rearrange)	D	A	B
PW (benefits)	4,700	6,480	11,250
PW (cost)	2,000	4,000	9,000

	A−D	B−A	
ΔB	1,780	4,770	
ΔC	2,000	5,000	
$\frac{\Delta B}{\Delta C}$ =	.89	.954	≤ 1.0

∴ choose D smaller cost

*** **12-8**

(a) The state engineer estimates that the cost of a canal is $680 million. The legislative analyst estimates the equivalent uniform annual cost of the investment for the canal to be $20.4 million. If the analyst expects the canal to last indefinitely, what interest rate is he using to compute the Equivalent Uniform Annual Cost (EUAC)?

(b) If the canal lasts only 50 years, what interest rate will the analyst be assuming if he believes the EUAC to be the same $20.4 million?

****

a) $A = P(A/P, i\%, n)$

For $n \to \infty$, $(A/P, i\%, n) = i$

$A = P \times i$

$i = A/P = \frac{20.4}{680} = 0.03$ or $\underline{i = 3\%}$

b) $A = P(A/P, i\%, n)$

$A = P(A/P, i\%, 50)$

$(A/P, i\%, 50) = A/P = \frac{20.4}{680} = 0.03$

$\underline{i = 1.75\%}$

12-9 **

The city council of Arson, Michigan is debating whether to buy a new fire truck to increase protection for the city. The financial analyst has prepared the following data:

	Truck A	Truck B
First cost	$50,000	$60,000
Maintenance	5,000/year	4,000/year
Useful life	7 years	7 years
Salvage	6,000	6,000
Reduction in fire damage	20,000/year	21,000/year

(a) Using the modified B/C ratio method determine whether the city should buy a new truck, and which one to buy if it will be paid for with money borrowed at an interest rate of 7%.

(b) How would the decision be affected if inflation were considered? Assume maintenance costs, salvage and fire damage are responsive to inflation.

(c) Is it possible the decision would change if the interest rate were lower? Why or why not?

****

a) In the modified B/C ratio, all annual cash flows belong in the numerator while first cost and salvage belong in the denominator. Either present or uniform equivalent methods may be used to relate cash flows.

$$\left(\frac{B}{C}\right)_A = \frac{20{,}000 - 5{,}000}{50{,}000(A/P, 7\%, 7) - 5{,}000(A/F, 7\%, 7)} = 1.72 \quad \therefore \text{ A is acceptable}$$

$$\left(\frac{\Delta B}{\Delta C}\right)_{B-A} = \frac{(21{,}000 - 20{,}000) - (4{,}000 - 5{,}000)}{10{,}000(A/P, 7\%, 7)} = 1.08 \quad \therefore \text{ B is better than A}$$

<u>Truck B should be purchased</u>.

b) Since both future cost (Maintenance) and benefits (reduced damage and salvage) are responsive to inflation, the decisions are not affected by inflation. Also note that a city government does not pay income tax.

c) No. Lower i (MARR) favors higher investment cost projects and truck B is already the highest cost alternative.

** **12-10**

The receiving area for the expansion of a meat processor is under study. The initial investment and annual operating benefits and costs are very different due to various degrees of automation. These can be summarized as follows:

	Alternative A	Alternative B	Alternative C
Investment costs	$180,000	$100,000	$280,000
Annual operating costs	16,000	12,000	28,000
Annual benefits	53,000	35,000	77,000

If the company's MARR is 10%, which alternative should be selected for a ten-year useful life analysis period. Use benefit/cost ratio analysis.

****

1. B/C ratios of individual alternatives.

$$(B/C)_A = \frac{EUAB_A}{EUAC_A} = \frac{53,000}{180,000\underbrace{(A/P,10\%,10)}_{0.1627}+16,000} = 1.17\ (>1)$$

$$(B/C)_B = \frac{EUAB_B}{EUAC_B} = \frac{35,000}{100,000(A/P,10\%,10)+12,000} = 1.24\ (>1)$$

$$(B/C)_C = \frac{EUAB_C}{EUAC_C} \quad \frac{77,000}{280,000(A/P,10\%,10)+28,000} = 1.047\ (>1)$$

∴ All economically attractive

2. Incremental B/C analysis.

Cost B < Cost A < Cost C

a) A-B : ΔBenefits = 53,000 - 35,000 = 18,000

$\Delta Costs = (180,000-100,000)(A/P,10\%,10)+(16,000-12,000)$
$= 17,016.$

$$\frac{\Delta Ben}{\Delta Costs} = \frac{\Delta EUAB}{\Delta EUAC} = \frac{18,000}{17,016} = 1.058\ (>1) \therefore \text{keep A}$$

b) C-A: ΔBenefits = 77,000 - 53,000 = 24,000

$\Delta Costs = (280K-180K)(A/P,10\%,10)+(28K-16K) = 28,270$

$$\frac{\Delta Ben}{\Delta Costs} = \frac{\Delta EUAB}{\Delta EUAC} = \frac{24,000}{28,270} = 0.8489\ (<1) \therefore \text{keep A}$$

Select Alternative A. (If we consider the operating costs as a reduction to the annual benefits a diff. numerical value for the B/C ratio might be found. The decision of which alt. is best will, however, be the same).

12-11 **

A firm is deciding a purchase between two machines.

Machine	Cost	Annual Savings	Useful Life	Salvage
X	$1000	$300	5 yrs	$0
Y	1200	325	5	0

With an assumed interest rate of 12%, which machine should be purchased based upon Benefit-Cost Analysis?

****

Machine X:

PW(cost) = $1,000

PW(benefits) = $300 (P/A, 12%, 5) = $300(3.605) = $1,081.5

B/C ratio = 1,081.5/1,000 = 1.0815

Machine Y:

PW (cost) = $1,200

PW (benefits) = $325 (P/A, 12%, 5) = 325(3.605) = 1,171.63

B/C ratio = 1,171.63/1,200 = .9763

Select Machine X

12-12 **

A public project has been analyzed using Benefit Cost ratios and been found to be acceptable. It was then found that the analyst should have used a lower interest rate. Using the correct interest rate, will the B/C ratio increase or decrease? Explain.

****

When the interest rate decreases, the present worth of future benefits increases relative to the investment costs. Therefore the B/C ratio increases.

13
OTHER PROBLEMS

** **13-1**

A small surveying company identifies its available independent alternatives as follows:

	Alternative	Initial Cost	Rate of Return
A.	Repair existing equipment	$1000	30%
B.	Buy EDM instrument	2500	9%
C.	Buy a new printer	3000	11%
D.	Buy equipment for an additional crew	3000	15%
E.	Buy programmable calculator	500	25%

The owner of the company has $5000 of savings currently invested in securities yielding 8% that could be used for the company.

(a) Assuming his funds are limited to his savings, what is the apparent cut-off rate of return?
(b) If he can borrow money at 10%, how much should be borrowed?

****

a)

Alt	Invest.	Cumm. Invest.	ROR
A	1,000	1,000	30%
E	500	1,500	25%
D	3,000	4,500	15%
C	3,000	7,500	11% ←
B	2,500	10,000	9%

cutoff rate-of-return = 11% to 15%

b) Do all projects with a rate of return > 10% Thus Alternatives A, E, D, & C with a total initial cost of $7,500 would be selected. Since only $5,000 is available, $2,500 would need to be borrowed.

13-2 **

The capital structure of a firm is given below.

Source of Capital	Percent of Capitalization	Interest Rate
Loans	35%	17%
Bonds	30	8
Common Stock	35	12

The combined income tax rate for the firm is 50%. What is the after-tax and before-tax cost of capital to the firm?

****

After Tax Cost of Capital:

$0.35 \times 17 + 0.30 \times 8 + 0.35 \times 12 = 12.55\%$

Before Tax Cost of Capital:

Common Stock must earn twice the 12% or 24%

$0.35 \times 17 + 0.30 \times 8 + 0.35 \times 24 = 16.75\%$

13-3 **

A local finance company will loan $10,000 to a homeowner. It is to be repaid in 2-1/2 years by making monthly payments of $500 each. What is the

(a) monthly interest rate?
(b) nominal interest rate?
(c) effective interest rate?

****

(a) $P = A(P/A, i\%, n)$

$10{,}000 = 500(P/A, i\%, 30)$

Trial & Error Solution:

i	$(P/A, i\%, 30)$
2%	22.396
3%	19.600

interpolate: $i = 2\% + (1\%)\left(\frac{22.396 - 20}{22.396 - 19.600}\right)$

$= 2\% + \frac{2.396}{2.796} = \underline{2.857\% \text{ per month}}$

(b) nominal interest rate $= 12i = 12(2.857) = \underline{34.28\%}$

(c) effective interest rate $= (1+i)^m - 1 = (1 + 0.02857)^{12} - 1$

$= \underline{40.22\%}$

** **13-4**

A small construction company identifies the following alternatives which are independent, except where noted.

Alternative	Initial Cost	Incremental Rate of Return	On Investment Over
1. Repair bulldozer	$ 5,000	30%	Ø
2. Replace backhoe			
With Model A	20,000	15%	Ø
With Model B	25,000	10.5%	2A
3. Buy new dump truck			
Model X	20,000	20%	Ø
Model Y	30,000	14%	3X
4. Buy computer			
Model K	5,000	12%	Ø
Model L	10,000	9.5%	4K

(a) Assuming the company has $55,000 available for investment and it is not able to borrow money, what alternatives should be chosen, and what is the cut-off rate of return?

(b) If the company can also borrow money at 10%, how much should be borrowed, and which alternatives should be selected?

****

Rank the alternatives by ΔROR

Project	Incremental Investment	Cummulative Investment	ΔROR
1	5,000	5,000	30%
3X	20,000	25,000	20%
2A	20,000	45,000	15%
3Y-3X	10,000	55,000	14%
4K	5,000	60,000	12%
2B-2A	5,000	65,000	10.5%
4L-4K	5,000	70,000	9.5%

a) With $55,000 available choose: 1 Repair bulldozer
2A Blackhoe model A
3Y Dump truck model Y

Cutoff Rate-of-Return = 12 to 14% No computer

b) Money borrowed = $10,000. Projects: 1 Repair bulldozer
2B Backhoe model B
3Y Dump truck model Y
4K Computer model K

13-5 **

The following independent and indivisible investment opportunities are available:

Investment	Amount	Rate of Return
A	$200	20%
B	100	22
C	50	19
D	100	18
E	50	15
F	Unlimited	7

(a) Which investments should be selected if the Minimum Attractive Rate of Return (MARR) is greater than or equal to 18%?

(b) Which investments should be selected if MARR is greater than or equal to 14% and the available budget is $400?

(c) What is the opportunity cost of capital in Part (b)?

****

a) A, B, C, D

b) A, B, D

c) 7%

13-6 **

The probability that a machine will last a certain number of years is given in the following table.

Years of Life	Probability of Obtaining Life
10	0.15
11	0.20
12	0.25
13	0.20
14	0.15
15	0.05

What is the expected life of the machine?

****

Expected value $= 10(0.15) + 11(0.20) + 12(0.25) + 13(0.2) + 14(0.15) + 15(0.05) = \underline{12.15 \text{ years}}$

** **13-7**

A city engineer calculated the present worth of benefits and costs of a number of possible projects, based on 10% interest cost and a 10 year analysis period.

Costs and Benefits in 1000's

Project:	A	B	C	D	E	F	G
Present Worth of Costs	75	70	50	35	60	25	70
Present Worth of Benefits	105	95	63	55	76	32	100

If 10% is a satisfactory Minimum Attractive Rate of Return (MARR), which project(s) should be selected if $180,000 is available for expenditure?

****

Project	A	B	C	D	E	F	G
Cost	75	70	50	35	60	25	70
Benefit	105	95	63	55	76	32	100
NPW	30	25	13	20	16	7	30
$\frac{NPW}{C}$	0.400	.357	.260	.571	.2667	.28	.428
	3	4	7	1	6	5	2

D + G + A = 35K + 70K + 75K = 180K

CHOOSE PROJECTS D, G, A

** **13-8**

How much would you pay now for an insurance policy protecting you aginst a one in twenty change of losing $10,000 three years from now if **i** = 10%?

****

$$PW = \frac{10{,}000}{20}(P/F, 10\%, 3) = 500(.7513) = \$375.65$$

13-9 **

A coal burning power plant has been ordered by the government to install a $5 million pollution abatement device to remove sulphur that is currently being emitted into the air. The sulphur is removed by allowing the plant's exhaust to pass through a filter. The filtration system requires the presence of a certain chemical. The purchase price of the chemical is $1000 per kilogram. Studies have been conducted that show that the number of units of sulphur that may be recovered annually from the exhaust is equal to 100 times the square root of the number of kilograms of the chemical used in the filtration system. Thus:

$$\text{(Units of Sulphur)} = 100 \times \text{(Kilograms of Chemical)}^{1/2}$$

Each unit of sulphur that is removed may then be sold by the power plant to chemical supply companies for $300. The filtration system and chemical have an expected life of 20 years at which time the chemical will have a resale value of $500 per kilogram, while the filtration system itself has no resale value.

Using a before-tax Minimum Attractive Rate of Return (MARR) of 10%, find the optimal amount of the chemical that should be purchased by the power plant.

****

Let x = no. of kilograms of chemicals purchased

Net Annual Cost (x) =

(purchase cost of pollution abatment device)(A/P, 10%, 20)

+ (chemical purchase cost)(A/P, 10%, 20)

- (salvage value of chemical)(A/F, 10%, 20)

- (annual sale value of sulphur)

$$\text{Net Annual Cost }(x) = (5{,}000{,}000)(A/P, 10\%, 20) + (1{,}000x)(A/P, 10\%, 20) - (500x)(A/F, 10\%, 20) - (300)(100\sqrt{x})$$

$$= 5{,}000(.1175) + 1{,}000x(.1175) - 500x(.0175) - 30{,}000\sqrt{x}$$

$$= 587.5 + 108.75x - 30{,}000\sqrt{x}$$

To minimize cost differentiate N.A.C.(x) and set = 0

$$\frac{d\,NAC(x)}{dx} = 108.75 - \frac{30{,}000}{2\sqrt{x}} = 0$$

$$\frac{1}{2\sqrt{x}} = \frac{108.75}{30{,}000} \rightarrow \sqrt{x} = \frac{30{,}000}{(2)(108.75)} = 137.931$$

$x = (137.931)^2 = \underline{19,024.97}$

check for max. or min. $\frac{d^2NAC(x)}{dx^2} = \frac{7,500}{\sqrt{x^3}} > 0$ for $x>0$

** **13-10**

How many positive rates of return does the following cash flow have?

Year	Cash Flow
0	-$50,000
1	+25,000
2	+25,000
3	0
4	-50,000
5	+25,000
6	+25,000
7	+25,000

****

Year	Accum. Cash Flow
0	-50,000
1	-25,000
2	0
3	0
4	-50,000
5	-25,000
6	0
7	+25,000 = A_7

3 sign changes in cash flow. Cash Flow Rule of Signs: There may be as many as 3 positive ROR.

Accum. Cash Flow: $A_7 \neq 0$, one sign change in accumulated cash flow, ∴ one positive rate of return

note: also check to see if there is any external investment during project life

13-11 ***

A bond with a face value of $1500 can be purchased for $800. The bond will mature five years from now and the bond dividend rate is 12%. If dividends are paid each three months, what effective interest rate would an investor receive if she purchases the bond?

****

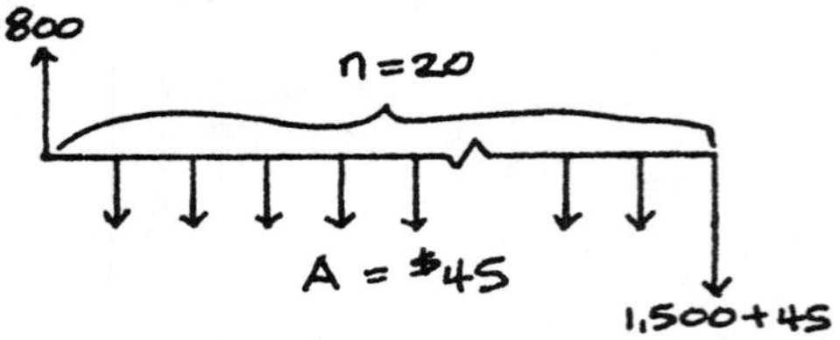

Interest period = 3 months

Compute i = interest rate / interest period

$0 = -800 + 45(P/A, i\%, 20) + 1{,}500(P/F, i\%, 20)$

Trial & Error Solution:

i	Net Present Worth
7%	$-800 + 45(10.594) + 1{,}500(0.2584) = +64.33$
8%	$-800 + 45(9.818) + 1{,}500(0.2145) = -36.44$

Linear interpolation: $i = 7\% + (1\%)\left(\frac{64.33}{64.33-(-36.44)}\right) = 7.638\%$

Effective interest rate $= (1+i)^m - 1 = (1+0.07638)^4 - 1$

$= 0.3423 = 34.23\%$

(Hand Calculator Solution: Effective interest rate = 34.14%)

13-12 ***

If the quarterly effective interest rate is 5-1/2%, with continuous compounding, what is the nominal interest rate?

****

$i_{quarterly} = 0.055$

$i_{effective} = e^{rt} - 1$

$0.055 = e^{rt} - 1$

$e^{1/4 r} = 1.055 \rightarrow r = 4\ln(1.055) = 4(0.053541) = \underline{21.42\%}$

Compound Interest Tables

Discrete End-Of-Period Compounding

Tables for:

1/4%	1/2%	3/4%	1%	1 1/4%	1 1/2%
1 3/4%	2%	2 1/2%	3%	3 1/2%	4%
4 1/2%	5%	6%	7%	8%	9%
10%	12%	15%	18%	20%	25%
30%	35%	40%	45%	50%	60%

Values Of Interest Factors When *N* Equals Infinity

Single Payment:

$(F/P,i,\infty) = \infty$

$(P/F,i,\infty) = 0$

Arithmetic Gradient Series:

$(A/G,i,\infty) = 1/i$

$(P/G,i,\infty) = 1/i^2$

Uniform Payment Series:

$(A/F,i,\infty) = 0$

$(A/P,i,\infty) = i$

$(F/A,i,\infty) = \infty$

$(P/A,i,\infty) = 1/i$

Compound Interest Factors

1/4%

	Single Payment		Uniform Payment Series				Arithmetic Gradient		
	Compound Amount Factor	Present Worth Factor	Sinking Fund Factor	Capital Recovery Factor	Compound Amount Factor	Present Worth Factor	Gradient Uniform Series	Gradient Present Worth	
n	Find F Given P F/P	Find P Given F P/F	Find A Given F A/F	Find A Given P A/P	Find F Given A F/A	Find P Given A P/A	Find A Given G A/G	Find P Given G P/G	n
1	1.003	.9975	1.0000	1.0025	1.000	0.998	0	0	1
2	1.005	.9950	.4994	.5019	2.003	1.993	0.504	1.005	2
3	1.008	.9925	.3325	.3350	3.008	2.985	1.005	2.999	3
4	1.010	.9901	.2491	.2516	4.015	3.975	1.501	5.966	4
5	1.013	.9876	.1990	.2015	5.025	4.963	1.998	9.916	5
6	1.015	.9851	.1656	.1681	6.038	5.948	2.498	14.861	6
7	1.018	.9827	.1418	.1443	7.053	6.931	2.995	20.755	7
8	1.020	.9802	.1239	.1264	8.070	7.911	3.490	27.611	8
9	1.023	.9778	.1100	.1125	9.091	8.889	3.987	35.440	9
10	1.025	.9753	.0989	.1014	10.113	9.864	4.483	44.216	10
11	1.028	.9729	.0898	.0923	11.139	10.837	4.978	53.950	11
12	1.030	.9705	.0822	.0847	12.167	11.807	5.474	64.634	12
13	1.033	.9681	.0758	.0783	13.197	12.775	5.968	76.244	13
14	1.036	.9656	.0703	.0728	14.230	13.741	6.464	88.826	14
15	1.038	.9632	.0655	.0680	15.266	14.704	6.957	102.301	15
16	1.041	.9608	.0613	.0638	16.304	15.665	7.451	116.716	16
17	1.043	.9584	.0577	.0602	17.344	16.624	7.944	132.063	17
18	1.046	.9561	.0544	.0569	18.388	17.580	8.437	148.319	18
19	1.049	.9537	.0515	.0540	19.434	18.533	8.929	165.492	19
20	1.051	.9513	.0488	.0513	20.482	19.485	9.421	183.559	20
21	1.054	.9489	.0464	.0489	21.534	20.434	9.912	202.531	21
22	1.056	.9465	.0443	.0468	22.587	21.380	10.404	222.435	22
23	1.059	.9442	.0423	.0448	23.644	22.324	10.894	243.212	23
24	1.062	.9418	.0405	.0430	24.703	23.266	11.384	264.854	24
25	1.064	.9395	.0388	.0413	25.765	24.206	11.874	287.407	25
26	1.067	.9371	.0373	.0398	26.829	25.143	12.363	310.848	26
27	1.070	.9348	.0358	.0383	27.896	26.078	12.852	335.150	27
28	1.072	.9325	.0345	.0370	28.966	27.010	13.341	360.343	28
29	1.075	.9301	.0333	.0358	30.038	27.940	13.828	386.366	29
30	1.078	.9278	.0321	.0346	31.114	28.868	14.317	413.302	30
36	1.094	.9140	.0266	.0291	37.621	34.387	17.234	592.632	36
40	1.105	.9049	.0238	.0263	42.014	38.020	19.171	728.882	40
48	1.127	.8871	.0196	.0221	50.932	45.179	23.025	1 040.22	48
50	1.133	.8826	.0188	.0213	53.189	46.947	23.984	1 125.96	50
52	1.139	.8782	.0180	.0205	55.458	48.705	24.941	1 214.76	52
60	1.162	.8609	.0155	.0180	64.647	55.653	28.755	1 600.31	60
70	1.191	.8396	.0131	.0156	76.395	64.144	33.485	2 147.87	70
72	1.197	.8355	.0127	.0152	78.780	65.817	34.426	2 265.81	72
80	1.221	.8189	.0113	.0138	88.440	72.427	38.173	2 764.74	80
84	1.233	.8108	.0107	.0132	93.343	75.682	40.037	3 030.06	84
90	1.252	.7987	.00992	.0124	100.789	80.504	42.820	3 447.19	90
96	1.271	.7869	.00923	.0117	108.349	85.255	45.588	3 886.62	96
100	1.284	.7790	.00881	.0113	113.451	88.383	47.425	4 191.60	100
104	1.297	.7713	.00843	.0109	118.605	91.480	49.256	4 505.93	104
120	1.349	.7411	.00716	.00966	139.743	103.563	56.512	5 852.52	120
240	1.821	.5492	.00305	.00555	328.306	180.312	107.590	19 399.75	240
360	2.457	.4070	.00172	.00422	582.745	237.191	152.894	36 264.96	360
480	3.315	.3016	.00108	.00358	926.074	279.343	192.673	53 821.93	480

Compound Interest Factors

	Single Payment		Uniform Payment Series				Arithmetic Gradient		
	Compound Amount Factor	Present Worth Factor	Sinking Fund Factor	Capital Recovery Factor	Compound Amount Factor	Present Worth Factor	Gradient Uniform Series	Gradient Present Worth	
n	Find *F* Given *P* *F/P*	Find *P* Given *F* *P/F*	Find *A* Given *F* *A/F*	Find *A* Given *P* *A/P*	Find *F* Given *A* *F/A*	Find *P* Given *A* *P/A*	Find *A* Given *G* *A/G*	Find *P* Given *G* *P/G*	**n**
1	1.005	.9950	1.0000	1.0050	1.000	0.995	0	0	**1**
2	1.010	.9901	.4988	.5038	2.005	1.985	0.499	0.991	**2**
3	1.015	.9851	.3317	.3367	3.015	2.970	0.996	2.959	**3**
4	1.020	.9802	.2481	.2531	4.030	3.951	1.494	5.903	**4**
5	1.025	.9754	.1980	.2030	5.050	4.926	1.990	9.803	**5**
6	1.030	.9705	.1646	.1696	6.076	5.896	2.486	14.660	**6**
7	1.036	.9657	.1407	.1457	7.106	6.862	2.980	20.448	**7**
8	1.041	.9609	.1228	.1278	8.141	7.823	3.474	27.178	**8**
9	1.046	.9561	.1089	.1139	9.182	8.779	3.967	34.825	**9**
10	1.051	.9513	.0978	.1028	10.228	9.730	4.459	43.389	**10**
11	1.056	.9466	.0887	.0937	11.279	10.677	4.950	52.855	**11**
12	1.062	.9419	.0811	.0861	12.336	11.619	5.441	63.218	**12**
13	1.067	.9372	.0746	.0796	13.397	12.556	5.931	74.465	**13**
14	1.072	.9326	.0691	.0741	14.464	13.489	6.419	86.590	**14**
15	1.078	.9279	.0644	.0694	15.537	14.417	6.907	99.574	**15**
16	1.083	.9233	.0602	.0652	16.614	15.340	7.394	113.427	**16**
17	1.088	.9187	.0565	.0615	17.697	16.259	7.880	128.125	**17**
18	1.094	.9141	.0532	.0582	18.786	17.173	8.366	143.668	**18**
19	1.099	.9096	.0503	.0553	19.880	18.082	8.850	160.037	**19**
20	1.105	.9051	.0477	.0527	20.979	18.987	9.334	177.237	**20**
21	1.110	.9006	.0453	.0503	22.084	19.888	9.817	195.245	**21**
22	1.116	.8961	.0431	.0481	23.194	20.784	10.300	214.070	**22**
23	1.122	.8916	.0411	.0461	24.310	21.676	10.781	233.680	**23**
24	1.127	.8872	.0393	.0443	25.432	22.563	11.261	254.088	**24**
25	1.133	.8828	.0377	.0427	26.559	23.446	11.741	275.273	**25**
26	1.138	.8784	.0361	.0411	27.692	24.324	12.220	297.233	**26**
27	1.144	.8740	.0347	.0397	28.830	25.198	12.698	319.955	**27**
28	1.150	.8697	.0334	.0384	29.975	26.068	13.175	343.439	**28**
29	1.156	.8653	.0321	.0371	31.124	26.933	13.651	367.672	**29**
30	1.161	.8610	.0310	.0360	32.280	27.794	14.127	392.640	**30**
36	1.197	.8356	.0254	.0304	39.336	32.871	16.962	557.564	**36**
40	1.221	.8191	.0226	.0276	44.159	36.172	18.836	681.341	**40**
48	1.270	.7871	.0185	.0235	54.098	42.580	22.544	959.928	**48**
50	1.283	.7793	.0177	.0227	56.645	44.143	23.463	1 035.70	**50**
52	1.296	.7716	.0169	.0219	59.218	45.690	24.378	1 113.82	**52**
60	1.349	.7414	.0143	.0193	69.770	51.726	28.007	1 448.65	**60**
70	1.418	.7053	.0120	.0170	83.566	58.939	32.468	1 913.65	**70**
72	1.432	.6983	.0116	.0166	86.409	60.340	33.351	2 012.35	**72**
80	1.490	.6710	.0102	.0152	98.068	65.802	36.848	2 424.65	**80**
84	1.520	.6577	.00961	.0146	104.074	68.453	38.576	2 640.67	**84**
90	1.567	.6383	.00883	.0138	113.311	72.331	41.145	2 976.08	**90**
96	1.614	.6195	.00814	.0131	122.829	76.095	43.685	3 324.19	**96**
100	1.647	.6073	.00773	.0127	129.334	78.543	45.361	3 562.80	**100**
104	1.680	.5953	.00735	.0124	135.970	80.942	47.025	3 806.29	**104**
120	1.819	.5496	.00610	.0111	163.880	90.074	53.551	4 823.52	**120**
240	3.310	.3021	.00216	.00716	462.041	139.581	96.113	13 415.56	**240**
360	6.023	.1660	.00100	.00600	1 004.5	166.792	128.324	21 403.32	**360**
480	10.957	.0913	.00050	.00550	1 991.5	181.748	151.795	27 588.37	**480**

Compound Interest Factors

3/4%

	Single Payment		Uniform Payment Series				Arithmetic Gradient		
	Compound Amount Factor	Present Worth Factor	Sinking Fund Factor	Capital Recovery Factor	Compound Amount Factor	Present Worth Factor	Gradient Uniform Series	Gradient Present Worth	
n	Find *F* Given *P* *F/P*	Find *P* Given *F* *P/F*	Find *A* Given *F* *A/F*	Find *A* Given *P* *A/P*	Find *F* Given *A* *F/A*	Find *P* Given *A* *P/A*	Find *A* Given *G* *A/G*	Find *P* Given *G* *P/G*	**n**
1	1.008	.9926	1.0000	1.0075	1.000	0.993	0	0	**1**
2	1.015	.9852	.4981	.5056	2.008	1.978	0.499	0.987	**2**
3	1.023	.9778	.3308	.3383	3.023	2.956	0.996	2.943	**3**
4	1.030	.9706	.2472	.2547	4.045	3.926	1.492	5.857	**4**
5	1.038	.9633	.1970	.2045	5.076	4.889	1.986	9.712	**5**
6	1.046	.9562	.1636	.1711	6.114	5.846	2.479	14.494	**6**
7	1.054	.9490	.1397	.1472	7.160	6.795	2.971	20.187	**7**
8	1.062	.9420	.1218	.1293	8.213	7.737	3.462	26.785	**8**
9	1.070	.9350	.1078	.1153	9.275	8.672	3.951	34.265	**9**
10	1.078	.9280	.0967	.1042	10.344	9.600	4.440	42.619	**10**
11	1.086	.9211	.0876	.0951	11.422	10.521	4.927	51.831	**11**
12	1.094	.9142	.0800	.0875	12.508	11.435	5.412	61.889	**12**
13	1.102	.9074	.0735	.0810	13.602	12.342	5.897	72.779	**13**
14	1.110	.9007	.0680	.0755	14.704	13.243	6.380	84.491	**14**
15	1.119	.8940	.0632	.0707	15.814	14.137	6.862	97.005	**15**
16	1.127	.8873	.0591	.0666	16.932	15.024	7.343	110.318	**16**
17	1.135	.8807	.0554	.0629	18.059	15.905	7.822	124.410	**17**
18	1.144	.8742	.0521	.0596	19.195	16.779	8.300	139.273	**18**
19	1.153	.8676	.0492	.0567	20.339	17.647	8.777	154.891	**19**
20	1.161	.8612	.0465	.0540	21.491	18.508	9.253	171.254	**20**
21	1.170	.8548	.0441	.0516	22.653	19.363	9.727	188.352	**21**
22	1.179	.8484	.0420	.0495	23.823	20.211	10.201	206.170	**22**
23	1.188	.8421	.0400	.0475	25.001	21.053	10.673	224.695	**23**
24	1.196	.8358	.0382	.0457	26.189	21.889	11.143	243.924	**24**
25	1.205	.8296	.0365	.0440	27.385	22.719	11.613	263.834	**25**
26	1.214	.8234	.0350	.0425	28.591	23.542	12.081	284.421	**26**
27	1.224	.8173	.0336	.0411	29.805	24.360	12.548	305.672	**27**
28	1.233	.8112	.0322	.0397	31.029	25.171	13.014	327.576	**28**
29	1.242	.8052	.0310	.0385	32.261	25.976	13.479	350.122	**29**
30	1.251	.7992	.0298	.0373	33.503	26.775	13.942	373.302	**30**
36	1.309	.7641	.0243	.0318	41.153	31.447	16.696	525.038	**36**
40	1.348	.7416	.0215	.0290	46.447	34.447	18.507	637.519	**40**
48	1.431	.6986	.0174	.0249	57.521	40.185	22.070	886.899	**48**
50	1.453	.6882	.0166	.0241	60.395	41.567	22.949	953.911	**50**
52	1.475	.6780	.0158	.0233	63.312	42.928	23.822	1 022.64	**52**
60	1.566	.6387	.0133	.0208	75.425	48.174	27.268	1 313.59	**60**
70	1.687	.5927	.0109	.0184	91.621	54.305	31.465	1 708.68	**70**
72	1.713	.5839	.0105	.0180	95.008	55.477	32.289	1 791.33	**72**
80	1.818	.5500	.00917	.0167	109.074	59.995	35.540	2 132.23	**80**
84	1.873	.5338	.00859	.0161	116.428	62.154	37.137	2 308.22	**84**
90	1.959	.5104	.00782	.0153	127.881	65.275	39.496	2 578.09	**90**
96	2.049	.4881	.00715	.0147	139.858	68.259	41.812	2 854.04	**96**
100	2.111	.4737	.00675	.0143	148.147	70.175	43.332	3 040.85	**100**
104	2.175	.4597	.00638	.0139	156.687	72.035	44.834	3 229.60	**104**
120	2.451	.4079	.00517	.0127	193.517	78.942	50.653	3 998.68	**120**
240	6.009	.1664	.00150	.00900	667.901	111.145	85.422	9 494.26	**240**
360	14.731	.0679	.00055	.00805	1 830.8	124.282	107.115	13 312.50	**360**
480	36.111	.0277	.00021	.00771	4 681.5	129.641	119.662	15 513.16	**480**

	Single Payment		Uniform Payment Series				Arithmetic Gradient		
	Compound Amount Factor	Present Worth Factor	Sinking Fund Factor	Capital Recovery Factor	Compound Amount Factor	Present Worth Factor	Gradient Uniform Series	Gradient Present Worth	
n	Find *F* Given *P* *F/P*	Find *P* Given *F* *P/F*	Find *A* Given *F* *A/F*	Find *A* Given *P* *A/P*	Find *F* Given *A* *F/A*	Find *P* Given *A* *P/A*	Find *A* Given *G* *A/G*	Find *P* Given *G* *P/G*	*n*
1	1.010	.9901	1.0000	1.0100	1.000	0.990	0	0	**1**
2	1.020	.9803	.4975	.5075	2.010	1.970	0.498	0.980	**2**
3	1.030	.9706	.3300	.3400	3.030	2.941	0.993	2.921	**3**
4	1.041	.9610	.2463	.2563	4.060	3.902	1.488	5.804	**4**
5	1.051	.9515	.1960	.2060	5.101	4.853	1.980	9.610	**5**
6	1.062	.9420	.1625	.1725	6.152	5.795	2.471	14.320	**6**
7	1.072	.9327	.1386	.1486	7.214	6.728	2.960	19.917	**7**
8	1.083	.9235	.1207	.1307	8.286	7.652	3.448	26.381	**8**
9	1.094	.9143	.1067	.1167	9.369	8.566	3.934	33.695	**9**
10	1.105	.9053	.0956	.1056	10.462	9.471	4.418	41.843	**10**
11	1.116	.8963	.0865	.0965	11.567	10.368	4.900	50.806	**11**
12	1.127	.8874	.0788	.0888	12.682	11.255	5.381	60.568	**12**
13	1.138	.8787	.0724	.0824	13.809	12.134	5.861	71.112	**13**
14	1.149	.8700	.0669	.0769	14.947	13.004	6.338	82.422	**14**
15	1.161	.8613	.0621	.0721	16.097	13.865	6.814	94.481	**15**
16	1.173	.8528	.0579	.0679	17.258	14.718	7.289	107.273	**16**
17	1.184	.8444	.0543	.0643	18.430	15.562	7.761	120.783	**17**
18	1.196	.8360	.0510	.0610	19.615	16.398	8.232	134.995	**18**
19	1.208	.8277	.0481	.0581	20.811	17.226	8.702	149.895	**19**
20	1.220	.8195	.0454	.0554	22.019	18.046	9.169	165.465	**20**
21	1.232	.8114	.0430	.0530	23.239	18.857	9.635	181.694	**21**
22	1.245	.8034	.0409	.0509	24.472	19.660	10.100	198.565	**22**
23	1.257	.7954	.0389	.0489	25.716	20.456	10.563	216.065	**23**
24	1.270	.7876	.0371	.0471	26.973	21.243	11.024	234.179	**24**
25	1.282	.7798	.0354	.0454	28.243	22.023	11.483	252.892	**25**
26	1.295	.7720	.0339	.0439	29.526	22.795	11.941	272.195	**26**
27	1.308	.7644	.0324	.0424	30.821	23.560	12.397	292.069	**27**
28	1.321	.7568	.0311	.0411	32.129	24.316	12.852	312.504	**28**
29	1.335	.7493	.0299	.0399	33.450	25.066	13.304	333.486	**29**
30	1.348	.7419	.0287	.0387	34.785	25.808	13.756	355.001	**30**
36	1.431	.6989	.0232	.0332	43.077	30.107	16.428	494.620	**36**
40	1.489	.6717	.0205	.0305	48.886	32.835	18.178	596.854	**40**
48	1.612	.6203	.0163	.0263	61.223	37.974	21.598	820.144	**48**
50	1.645	.6080	.0155	.0255	64.463	39.196	22.436	879.417	**50**
52	1.678	.5961	.0148	.0248	67.769	40.394	23.269	939.916	**52**
60	1.817	.5504	.0122	.0222	81.670	44.955	26.533	1 192.80	**60**
70	2.007	.4983	.00993	.0199	100.676	50.168	30.470	1 528.64	**70**
72	2.047	.4885	.00955	.0196	104.710	51.150	31.239	1 597.86	**72**
80	2.217	.4511	.00822	.0182	121.671	54.888	34.249	1 879.87	**80**
84	2.307	.4335	.00765	.0177	130.672	56.648	35.717	2 023.31	**84**
90	2.449	.4084	.00690	.0169	144.863	59.161	37.872	2 240.56	**90**
96	2.599	.3847	.00625	.0163	159.927	61.528	39.973	2 459.42	**96**
100	2.705	.3697	.00587	.0159	170.481	63.029	41.343	2 605.77	**100**
104	2.815	.3553	.00551	.0155	181.464	64.471	42.688	2 752.17	**104**
120	3.300	.3030	.00435	.0143	230.039	69.701	47.835	3 334.11	**120**
240	10.893	.0918	00101	.0110	989.254	90.819	75.739	6 878.59	**240**
360	35.950	.0278	.00029	.0103	3 495.0	97.218	89.699	8 720.43	**360**
480	118.648	.00843	.00008	.0101	11 764.8	99.157	95.920	9 511.15	**480**

Compound Interest Factors

1¼%

	Single Payment		Uniform Payment Series				Arithmetic Gradient		
	Compound Amount Factor	Present Worth Factor	Sinking Fund Factor	Capital Recovery Factor	Compound Amount Factor	Present Worth Factor	Gradient Uniform Series	Gradient Present Worth	
n	Find *F* Given *P* *F/P*	Find *P* Given *F* *P/F*	Find *A* Given *F* *A/F*	Find *A* Given *P* *A/P*	Find *F* Given *A* *F/A*	Find *P* Given *A* *P/A*	Find *A* Given *G* *A/G*	Find *P* Given *G* *P/G*	n
1	1.013	.9877	1.0000	1.0125	1.000	0.988	0	0	**1**
2	1.025	.9755	.4969	.5094	2.013	1.963	0.497	0.976	**2**
3	1.038	.9634	.3292	.3417	3.038	2.927	0.992	2.904	**3**
4	1.051	.9515	.2454	.2579	4.076	3.878	1.485	5.759	**4**
5	1.064	.9398	.1951	.2076	5.127	4.818	1.976	9.518	**5**
6	1.077	.9282	.1615	.1740	6.191	5.746	2.464	14.160	**6**
7	1.091	.9167	.1376	.1501	7.268	6.663	2.951	19.660	**7**
8	1.104	.9054	.1196	.1321	8.359	7.568	3.435	25.998	**8**
9	1.118	.8942	.1057	.1182	9.463	8.462	3.918	33.152	**9**
10	1.132	.8832	.0945	.1070	10.582	9.346	4.398	41.101	**10**
11	1.146	.8723	.0854	.0979	11.714	10.218	4.876	49.825	**11**
12	1.161	.8615	.0778	.0903	12.860	11.079	5.352	59.302	**12**
13	1.175	.8509	.0713	.0838	14.021	11.930	5.827	69.513	**13**
14	1.190	.8404	.0658	.0783	15.196	12.771	6.299	80.438	**14**
15	1.205	.8300	.0610	.0735	16.386	13.601	6.769	92.058	**15**
16	1.220	.8197	.0568	.0693	17.591	14.420	7.237	104.355	**16**
17	1.235	.8096	.0532	.0657	18.811	15.230	7.702	117.309	**17**
18	1.251	.7996	.0499	.0624	20.046	16.030	8.166	130.903	**18**
19	1.266	.7898	.0470	.0595	21.297	16.819	8.628	145.119	**19**
20	1.282	.7800	.0443	.0568	22.563	17.599	9.088	159.940	**20**
21	1.298	.7704	.0419	.0544	23.845	18.370	9.545	175.348	**21**
22	1.314	.7609	.0398	.0523	25.143	19.131	10.001	191.327	**22**
23	1.331	.7515	.0378	.0503	26.458	19.882	10.455	207.859	**23**
24	1.347	.7422	.0360	.0485	27.788	20.624	10.906	224.930	**24**
25	1.364	.7330	.0343	.0468	29.136	21.357	11.355	242.523	**25**
26	1.381	.7240	.0328	.0453	30.500	22.081	11.803	260.623	**26**
27	1.399	.7150	.0314	.0439	31.881	22.796	12.248	279.215	**27**
28	1.416	.7062	.0300	.0425	33.280	23.503	12.691	298.284	**28**
29	1.434	.6975	.0288	.0413	34.696	24.200	13.133	317.814	**29**
30	1.452	.6889	.0277	.0402	36.129	24.889	13.572	337.792	**30**
36	1.564	.6394	.0222	.0347	45.116	28.847	16.164	466.297	**36**
40	1.644	.6084	.0194	.0319	51.490	31.327	17.852	559.247	**40**
48	1.815	.5509	.0153	.0278	65.229	35.932	21.130	759.248	**48**
50	1.861	.5373	.0145	.0270	68.882	37.013	21.930	811.692	**50**
52	1.908	.5242	.0138	.0263	72.628	38.068	22.722	864.960	**52**
60	2.107	.4746	.0113	.0238	88.575	42.035	25.809	1 084.86	**60**
70	2.386	.4191	.00902	.0215	110.873	46.470	29.492	1 370.47	**70**
72	2.446	.4088	.00864	.0211	115.675	47.293	30.205	1 428.48	**72**
80	2.701	.3702	.00735	.0198	136.120	50.387	32.983	1 661.89	**80**
84	2.839	.3522	.00680	.0193	147.130	51.822	34.326	1 778.86	**84**
90	3.059	.3269	.00607	.0186	164.706	53.846	36.286	1 953.85	**90**
96	3.296	.3034	.00545	.0179	183.643	55.725	38.180	2 127.55	**96**
100	3.463	.2887	.00507	.0176	197.074	56.901	39.406	2 242.26	**100**
104	3.640	.2747	.00474	.0172	211.190	58.021	40.604	2 355.90	**104**
120	4.440	.2252	.00363	.0161	275.220	61.983	45.119	2 796.59	**120**
240	19.716	.0507	.00067	.0132	1 497.3	75.942	67.177	5 101.55	**240**
360	87.543	.0114	.00014	.0126	6 923.4	79.086	75.840	5 997.91	**360**
480	388.713	.00257	.00003	.0125	31 017.1	79.794	78.762	6 284.74	**480**

Compound Interest Factors

1½%

	Single Payment		Uniform Payment Series				Arithmetic Gradient		
	Compound Amount Factor	Present Worth Factor	Sinking Fund Factor	Capital Recovery Factor	Compound Amount Factor	Present Worth Factor	Gradient Uniform Series	Gradient Present Worth	
n	Find F Given P F/P	Find P Given F P/F	Find A Given F A/F	Find A Given P A/P	Find F Given A F/A	Find P Given A P/A	Find A Given G A/G	Find P Given G P/G	n
1	1.015	.9852	1.0000	1.0150	1.000	0.985	0	0	**1**
2	1.030	.9707	.4963	.5113	2.015	1.956	0.496	0.970	**2**
3	1.046	.9563	.3284	.3434	3.045	2.912	0.990	2.883	**3**
4	1.061	.9422	.2444	.2594	4.091	3.854	1.481	5.709	**4**
5	1.077	.9283	.1941	.2091	5.152	4.783	1.970	9.422	**5**
6	1.093	.9145	.1605	.1755	6.230	5.697	2.456	13.994	**6**
7	1.110	.9010	.1366	.1516	7.323	6.598	2.940	19.400	**7**
8	1.126	.8877	.1186	.1336	8.433	7.486	3.422	25.614	**8**
9	1.143	.8746	.1046	.1196	9.559	8.360	3.901	32.610	**9**
10	1.161	.8617	.0934	.1084	10.703	9.222	4.377	40.365	**10**
11	1.178	.8489	.0843	.0993	11.863	10.071	4.851	48.855	**11**
12	1.196	.8364	.0767	.0917	13.041	10.907	5.322	58.054	**12**
13	1.214	.8240	.0702	.0852	14.237	11.731	5.791	67.943	**13**
14	1.232	.8118	.0647	.0797	15.450	12.543	6.258	78.496	**14**
15	1.250	.7999	.0599	.0749	16.682	13.343	6.722	89.694	**15**
16	1.269	.7880	.0558	.0708	17.932	14.131	7.184	101.514	**16**
17	1.288	.7764	.0521	.0671	19.201	14.908	7.643	113.937	**17**
18	1.307	.7649	.0488	.0638	20.489	15.673	8.100	126.940	**18**
19	1.327	.7536	.0459	.0609	21.797	16.426	8.554	140.505	**19**
20	1.347	.7425	.0432	.0582	23.124	17.169	9.005	154.611	**20**
21	1.367	.7315	.0409	.0559	24.470	17.900	9.455	169.241	**21**
22	1.388	.7207	.0387	.0537	25.837	18.621	9.902	184.375	**22**
23	1.408	.7100	.0367	.0517	27.225	19.331	10.346	199.996	**23**
24	1.430	.6995	.0349	.0499	28.633	20.030	10.788	216.085	**24**
25	1.451	.6892	.0333	.0483	30.063	20.720	11.227	232.626	**25**
26	1.473	.6790	.0317	.0467	31.514	21.399	11.664	249.601	**26**
27	1.495	.6690	.0303	.0453	32.987	22.068	12.099	266.995	**27**
28	1.517	.6591	.0290	.0440	34.481	22.727	12.531	284.790	**28**
29	1.540	.6494	.0278	.0428	35.999	23.376	12.961	302.972	**29**
30	1.563	.6398	.0266	.0416	37.539	24.016	13.388	321.525	**30**
36	1.709	.5851	.0212	.0362	47.276	27.661	15.901	439.823	**36**
40	1.814	.5513	.0184	.0334	54.268	29.916	17.528	524.349	**40**
48	2.043	.4894	.0144	.0294	69.565	34.042	20.666	703.537	**48**
50	2.105	.4750	.0136	.0286	73.682	35.000	21.428	749.955	**50**
52	2.169	.4611	.0128	.0278	77.925	35.929	22.179	796.868	**52**
60	2.443	.4093	.0104	.0254	96.214	39.380	25.093	988.157	**60**
70	2.835	.3527	.00817	.0232	122.363	43.155	28.529	1 231.15	**70**
72	2.921	.3423	.00781	.0228	128.076	43.845	29.189	1 279.78	**72**
80	3.291	.3039	.00655	.0215	152.710	46.407	31.742	1 473.06	**80**
84	3.493	.2863	.00602	.0210	166.172	47.579	32.967	1 568.50	**84**
90	3.819	.2619	.00532	.0203	187.929	49.210	34.740	1 709.53	**90**
96	4.176	.2395	.00472	.0197	211.719	50.702	36.438	1 847.46	**96**
100	4.432	.2256	.00437	.0194	228.802	51.625	37.529	1 937.43	**100**
104	4.704	.2126	.00405	.0190	246.932	52.494	38.589	2 025.69	**104**
120	5.969	.1675	.00302	.0180	331.286	55.498	42.518	2 359.69	**120**
240	35.632	.0281	.00043	.0154	2 308.8	64.796	59.737	3 870.68	**240**
360	212.700	.00470	.00007	.0151	14 113.3	66.353	64.966	4 310.71	**360**
480	1 269.7	.00079	.00001	.0150	84 577.8	66.614	66.288	4 415.74	**480**

1¾% Compound Interest Factors 1¾%

	Single Payment		Uniform Payment Series				Arithmetic Gradient		
	Compound Amount Factor	Present Worth Factor	Sinking Fund Factor	Capital Recovery Factor	Compound Amount Factor	Present Worth Factor	Gradient Uniform Series	Gradient Present Worth	
n	Find *F* Given *P* *F/P*	Find *P* Given *F* *P/F*	Find *A* Given *F* *A/F*	Find *A* Given *P* *A/P*	Find *F* Given *A* *F/A*	Find *P* Given *A* *P/A*	Find *A* Given *G* *A/G*	Find *P* Given *G* *P/G*	**n**
1	1.018	.9828	1.0000	1.0175	1.000	0.983	0	0	**1**
2	1.035	.9659	.4957	.5132	2.018	1.949	0.496	0.966	**2**
3	1.053	.9493	.3276	.3451	3.053	2.898	0.989	2.865	**3**
4	1.072	.9330	.2435	.2610	4.106	3.831	1.478	5.664	**4**
5	1.091	.9169	.1931	.2106	5.178	4.748	1.965	9.332	**5**
6	1.110	.9011	.1595	.1770	6.269	5.649	2.450	13.837	**6**
7	1.129	.8856	.1355	.1530	7.378	6.535	2.931	19.152	**7**
8	1.149	.8704	.1175	.1350	8.508	7.405	3.409	25.245	**8**
9	1.169	.8554	.1036	.1211	9.656	8.261	3.885	32.088	**9**
10	1.189	.8407	.0924	.1099	10.825	9.101	4.357	39.655	**10**
11	1.210	.8263	.0832	.1007	12.015	9.928	4.827	47.918	**11**
12	1.231	.8121	.0756	.0931	13.225	10.740	5.294	56.851	**12**
13	1.253	.7981	.0692	.0867	14.457	11.538	5.758	66.428	**13**
14	1.275	.7844	.0637	.0812	15.710	12.322	6.219	76.625	**14**
15	1.297	.7709	.0589	.0764	16.985	13.093	6.677	87.417	**15**
16	1.320	.7576	.0547	.0722	18.282	13.851	7.132	98.782	**16**
17	1.343	.7446	.0510	.0685	19.602	14.595	7.584	110.695	**17**
18	1.367	.7318	.0477	.0652	20.945	15.327	8.034	123.136	**18**
19	1.390	.7192	.0448	.0623	22.311	16.046	8.481	136.081	**19**
20	1.415	.7068	.0422	.0597	23.702	16.753	8.924	149.511	**20**
21	1.440	.6947	.0398	.0573	25.116	17.448	9.365	163.405	**21**
22	1.465	.6827	.0377	.0552	26.556	18.130	9.804	177.742	**22**
23	1.490	.6710	.0357	.0532	28.021	18.801	10.239	192.503	**23**
24	1.516	.6594	.0339	.0514	29.511	19.461	10.671	207.671	**24**
25	1.543	.6481	.0322	.0497	31.028	20.109	11.101	223.225	**25**
26	1.570	.6369	.0307	.0482	32.571	20.746	11.528	239.149	**26**
27	1.597	.6260	.0293	.0468	34.141	21.372	11.952	255.425	**27**
28	1.625	.6152	.0280	.0455	35.738	21.987	12.373	272.036	**28**
29	1.654	.6046	.0268	.0443	37.363	22.592	12.791	288.967	**29**
30	1.683	.5942	.0256	.0431	39.017	23.186	13.206	306.200	**30**
36	1.867	.5355	.0202	.0377	49.566	26.543	15.640	415.130	**36**
40	2.002	.4996	.0175	.0350	57.234	28.594	17.207	492.017	**40**
48	2.300	.4349	.0135	.0310	74.263	32.294	20.209	652.612	**48**
50	2.381	.4200	.0127	.0302	78.903	33.141	20.932	693.708	**50**
52	2.465	.4057	.0119	.0294	83.706	33.960	21.644	735.039	**52**
60	2.832	.3531	.00955	.0271	104.676	36.964	24.389	901.503	**60**
70	3.368	.2969	.00739	.0249	135.331	40.178	27.586	1 108.34	**70**
72	3.487	.2868	.00704	.0245	142.127	40.757	28.195	1 149.12	**72**
80	4.006	.2496	.00582	.0233	171.795	42.880	30.533	1 309.25	**80**
84	4.294	.2329	.00531	.0228	188.246	43.836	31.644	1 387.16	**84**
90	4.765	.2098	.00465	.0221	215.166	45.152	33.241	1 500.88	**90**
96	5.288	.1891	.00408	.0216	245.039	46.337	34.756	1 610.48	**96**
100	5.668	.1764	.00375	.0212	266.753	47.062	35.721	1 681.09	**100**
104	6.075	.1646	.00345	.0209	290.028	47.737	36.652	1 749.68	**104**
120	8.019	.1247	.00249	.0200	401.099	50.017	40.047	2 003.03	**120**
240	64.308	.0156	.00028	.0178	3 617.6	56.254	53.352	3 001.27	**240**
360	515.702	.00194	.00003	.0175	29 411.5	57.032	56.443	3 219.08	**360**
480	4 135.5	.00024		.0175	236 259.0	57.129	57.027	3 257.88	**480**

Compound Interest Factors

2%

	Single Payment		Uniform Payment Series				Arithmetic Gradient		
	Compound Amount Factor	Present Worth Factor	Sinking Fund Factor	Capital Recovery Factor	Compound Amount Factor	Present Worth Factor	Gradient Uniform Series	Gradient Present Worth	
n	Find *F* Given *P* *F/P*	Find *P* Given *F* *P/F*	Find *A* Given *F* *A/F*	Find *A* Given *P* *A/P*	Find *F* Given *A* *F/A*	Find *P* Given *A* *P/A*	Find *A* Given *G* *A/G*	Find *P* Given *G* *P/G*	n
1	1.020	.9804	1.0000	1.0200	1.000	0.980	0	0	**1**
2	1.040	.9612	.4951	.5151	2.020	1.942	0.495	0.961	**2**
3	1.061	.9423	.3268	.3468	3.060	2.884	0.987	2.846	**3**
4	1.082	.9238	.2426	.2626	4.122	3.808	1.475	5.617	**4**
5	1.104	.9057	.1922	.2122	5.204	4.713	1.960	9.240	**5**
6	1.126	.8880	.1585	.1785	6.308	5.601	2.442	13.679	**6**
7	1.149	.8706	.1345	.1545	7.434	6.472	2.921	18.903	**7**
8	1.172	.8535	.1165	.1365	8.583	7.325	3.396	24.877	**8**
9	1.195	.8368	.1025	.1225	9.755	8.162	3.868	31.571	**9**
10	1.219	.8203	.0913	.1113	10.950	8.983	4.337	38.954	**10**
11	1.243	.8043	.0822	.1022	12.169	9.787	4.802	46.996	**11**
12	1.268	.7885	.0746	.0946	13.412	10.575	5.264	55.669	**12**
13	1.294	.7730	.0681	.0881	14.680	11.348	5.723	64.946	**13**
14	1.319	.7579	.0626	.0826	15.974	12.106	6.178	74.798	**14**
15	1.346	.7430	.0578	.0778	17.293	12.849	6.631	85.200	**15**
16	1.373	.7284	.0537	.0737	18.639	13.578	7.080	96.127	**16**
17	1.400	.7142	.0500	.0700	20.012	14.292	7.526	107.553	**17**
18	1.428	.7002	.0467	.0667	21.412	14.992	7.968	119.456	**18**
19	1.457	.6864	.0438	.0638	22.840	15.678	8.407	131.812	**19**
20	1.486	.6730	.0412	.0612	24.297	16.351	8.843	144.598	**20**
21	1.516	.6598	.0388	.0588	25.783	17.011	9.276	157.793	**21**
22	1.546	.6468	.0366	.0566	27.299	17.658	9.705	171.377	**22**
23	1.577	.6342	.0347	.0547	28.845	18.292	10.132	185.328	**23**
24	1.608	.6217	.0329	.0529	30.422	18.914	10.555	199.628	**24**
25	1.641	.6095	.0312	.0512	32.030	19.523	10.974	214.256	**25**
26	1.673	.5976	.0297	.0497	33.671	20.121	11.391	229.196	**26**
27	1.707	.5859	.0283	.0483	35.344	20.707	11.804	244.428	**27**
28	1.741	.5744	.0270	.0470	37.051	21.281	12.214	259.936	**28**
29	1.776	.5631	.0258	.0458	38.792	21.844	12.621	275.703	**29**
30	1.811	.5521	.0247	.0447	40.568	22.396	13.025	291.713	**30**
36	2.040	.4902	.0192	.0392	51.994	25.489	15.381	392.036	**36**
40	2.208	.4529	.0166	.0366	60.402	27.355	16.888	461.989	**40**
48	2.587	.3865	.0126	.0326	79.353	30.673	19.755	605.961	**48**
50	2.692	.3715	.0118	.0318	84.579	31.424	20.442	642.355	**50**
52	2.800	.3571	.0111	.0311	90.016	32.145	21.116	678.779	**52**
60	3.281	.3048	.00877	.0288	114.051	34.761	23.696	823.692	**60**
70	4.000	.2500	.00667	.0267	149.977	37.499	26.663	999.829	**70**
72	4.161	.2403	.00633	.0263	158.056	37.984	27.223	1 034.050	**72**
80	4.875	.2051	.00516	.0252	193.771	39.744	29.357	1 166.781	**80**
84	5.277	.1895	.00468	.0247	213.865	40.525	30.361	1 230.413	**84**
90	5.943	.1683	.00405	.0240	247.155	41.587	31.793	1 322.164	**90**
96	6.693	.1494	.00351	.0235	284.645	42.529	33.137	1 409.291	**96**
100	7.245	.1380	.00320	.0232	312.230	43.098	33.986	1 464.747	**100**
104	7.842	.1275	.00292	.0229	342.090	43.624	34.799	1 518.082	**104**
120	10.765	.0929	.00205	.0220	488.255	45.355	37.711	1 710.411	**120**
240	115.887	.00863	.00017	.0202	5 744.4	49.569	47.911	2 374.878	**240**
360	1 247.5	.00080	.00002	.0200	62 326.8	49.960	49.711	2 483.567	**360**
480	13 429.8	.00007		.0200	671 442.0	49.996	49.964	2 498.027	**480**

Compound Interest Factors

2½%

	Single Payment		Uniform Payment Series				Arithmetic Gradient		
	Compound Amount Factor	Present Worth Factor	Sinking Fund Factor	Capital Recovery Factor	Compound Amount Factor	Present Worth Factor	Gradient Uniform Series	Gradient Present Worth	
n	Find *F* Given *P* *F/P*	Find *P* Given *F* *P/F*	Find *A* Given *F* *A/F*	Find *A* Given *P* *A/P*	Find *F* Given *A* *F/A*	Find *P* Given *A* *P/A*	Find *A* Given *G* *A/G*	Find *P* Given *G* *P/G*	n
1	1.025	.9756	1.0000	1.0250	1.000	0.976	0	0	**1**
2	1.051	.9518	.4938	.5188	2.025	1.927	0.494	0.952	**2**
3	1.077	.9286	.3251	.3501	3.076	2.856	0.984	2.809	**3**
4	1.104	.9060	.2408	.2658	4.153	3.762	1.469	5.527	**4**
5	1.131	.8839	.1902	.2152	5.256	4.646	1.951	9.062	**5**
6	1.160	.8623	.1566	.1816	6.388	5.508	2.428	13.374	**6**
7	1.189	.8413	.1325	.1575	7.547	6.349	2.901	18.421	**7**
8	1.218	.8207	.1145	.1395	8.736	7.170	3.370	24.166	**8**
9	1.249	.8007	.1005	.1255	9.955	7.971	3.835	30.572	**9**
10	1.280	.7812	.0893	.1143	11.203	8.752	4.296	37.603	**10**
11	1.312	.7621	.0801	.1051	12.483	9.514	4.753	45.224	**11**
12	1.345	.7436	.0725	.0975	13.796	10.258	5.206	53.403	**12**
13	1.379	.7254	.0660	.0910	15.140	10.983	5.655	62.108	**13**
14	1.413	.7077	.0605	.0855	16.519	11.691	6.100	71.309	**14**
15	1.448	.6905	.0558	.0808	17.932	12.381	6.540	80.975	**15**
16	1.485	.6736	.0516	.0766	19.380	13.055	6.977	91.080	**16**
17	1.522	.6572	.0479	.0729	20.865	13.712	7.409	101.595	**17**
18	1.560	.6412	.0447	.0697	22.386	14.353	7.838	112.495	**18**
19	1.599	.6255	.0418	.0668	23.946	14.979	8.262	123.754	**19**
20	1.639	.6103	.0391	.0641	25.545	15.589	8.682	135.349	**20**
21	1.680	.5954	.0368	.0618	27.183	16.185	9.099	147.257	**21**
22	1.722	.5809	.0346	.0596	28.863	16.765	9.511	159.455	**22**
23	1.765	.5667	.0327	.0577	30.584	17.332	9.919	171.922	**23**
24	1.809	.5529	.0309	.0559	32.349	17.885	10.324	184.638	**24**
25	1.854	.5394	.0293	.0543	34.158	18.424	10.724	197.584	**25**
26	1.900	.5262	.0278	.0528	36.012	18.951	11.120	210.740	**26**
27	1.948	.5134	.0264	.0514	37.912	19.464	11.513	224.088	**27**
28	1.996	.5009	.0251	.0501	39.860	19.965	11.901	237.612	**28**
29	2.046	.4887	.0239	.0489	41.856	20.454	12.286	251.294	**29**
30	2.098	.4767	.0228	.0478	43.903	20.930	12.667	265.120	**30**
31	2.150	.4651	.0217	.0467	46.000	21.395	13.044	279.073	**31**
32	2.204	.4538	.0208	.0458	48.150	21.849	13.417	293.140	**32**
33	2.259	.4427	.0199	.0449	50.354	22.292	13.786	307.306	**33**
34	2.315	.4319	.0190	.0440	52.613	22.724	14.151	321.559	**34**
35	2.373	.4214	.0182	.0432	54.928	23.145	14.512	335.886	**35**
40	2.685	.3724	.0148	.0398	67.402	25.103	16.262	408.221	**40**
45	3.038	.3292	.0123	.0373	81.516	26.833	17.918	480.806	**45**
50	3.437	.2909	.0103	.0353	97.484	28.362	19.484	552.607	**50**
55	3.889	.2572	.00865	.0337	115.551	29.714	20.961	622.827	**55**
60	4.400	.2273	.00735	.0324	135.991	30.909	22.352	690.865	**60**
65	4.978	.2009	.00628	.0313	159.118	31.965	23.660	756.280	**65**
70	5.632	.1776	.00540	.0304	185.284	32.898	24.888	818.763	**70**
75	6.372	.1569	.00465	.0297	214.888	33.723	26.039	878.114	**75**
80	7.210	.1387	.00403	.0290	248.382	34.452	27.117	934.217	**80**
85	8.157	.1226	.00349	.0285	286.278	35.096	28.123	987.026	**85**
90	9.229	.1084	.00304	.0280	329.154	35.666	29.063	1 036.54	**90**
95	10.442	.0958	.00265	.0276	377.663	36.169	29.938	1 082.83	**95**
100	11.814	.0846	.00231	.0273	432.548	36.614	30.752	1 125.97	**100**

Compound Interest Factors

3%

	Single Payment		Uniform Payment Series				Arithmetic Gradient		
	Compound Amount Factor	Present Worth Factor	Sinking Fund Factor	Capital Recovery Factor	Compound Amount Factor	Present Worth Factor	Gradient Uniform Series	Gradient Present Worth	
n	Find *F* Given *P* *F/P*	Find *P* Given *F* *P/F*	Find *A* Given *F* *A/F*	Find *A* Given *P* *A/P*	Find *F* Given *A* *F/A*	Find *P* Given *A* *P/A*	Find *A* Given *G* *A/G*	Find *P* Given *G* *P/G*	*n*
1	1.030	.9709	1.0000	1.0300	1.000	0.971	0	0	**1**
2	1.061	.9426	.4926	.5226	2.030	1.913	0.493	0.943	**2**
3	1.093	.9151	.3235	.3535	3.091	2.829	0.980	2.773	**3**
4	1.126	.8885	.2390	.2690	4.184	3.717	1.463	5.438	**4**
5	1.159	.8626	.1884	.2184	5.309	4.580	1.941	8.889	**5**
6	1.194	.8375	.1546	.1846	6.468	5.417	2.414	13.076	**6**
7	1.230	.8131	.1305	.1605	7.662	6.230	2.882	17.955	**7**
8	1.267	.7894	.1125	.1425	8.892	7.020	3.345	23.481	**8**
9	1.305	.7664	.0984	.1284	10.159	7.786	3.803	29.612	**9**
10	1.344	.7441	.0872	.1172	11.464	8.530	4.256	36.309	**10**
11	1.384	.7224	.0781	.1081	12.808	9.253	4.705	43.533	**11**
12	1.426	.7014	.0705	.1005	14.192	9.954	5.148	51.248	**12**
13	1.469	.6810	.0640	.0940	15.618	10.635	5.587	59.419	**13**
14	1.513	.6611	.0585	.0885	17.086	11.296	6.021	68.014	**14**
15	1.558	.6419	.0538	.0838	18.599	11.938	6.450	77.000	**15**
16	1.605	.6232	.0496	.0796	20.157	12.561	6.874	86.348	**16**
17	1.653	.6050	.0460	.0760	21.762	13.166	7.294	96.028	**17**
18	1.702	.5874	.0427	.0727	23.414	13.754	7.708	106.014	**18**
19	1.754	.5703	.0398	.0698	25.117	14.324	8.118	116.279	**19**
20	1.806	.5537	.0372	.0672	26.870	14.877	8.523	126.799	**20**
21	1.860	.5375	.0349	.0649	28.676	15.415	8.923	137.549	**21**
22	1.916	.5219	.0327	.0627	30.537	15.937	9.319	148.509	**22**
23	1.974	.5067	.0308	.0608	32.453	16.444	9.709	159.656	**23**
24	2.033	.4919	.0290	.0590	34.426	16.936	10.095	170.971	**24**
25	2.094	.4776	.0274	.0574	36.459	17.413	10.477	182.433	**25**
26	2.157	.4637	.0259	.0559	38.553	17.877	10.853	194.026	**26**
27	2.221	.4502	.0246	.0546	40.710	18.327	11.226	205.731	**27**
28	2.288	.4371	.0233	.0533	42.931	18.764	11.593	217.532	**28**
29	2.357	.4243	.0221	.0521	45.219	19.188	11.956	229.413	**29**
30	2.427	.4120	.0210	.0510	47.575	19.600	12.314	241.361	**30**
31	2.500	.4000	.0200	.0500	50.003	20.000	12.668	253.361	**31**
32	2.575	.3883	.0190	.0490	52.503	20.389	13.017	265.399	**32**
33	2.652	.3770	.0182	.0482	55.078	20.766	13.362	277.464	**33**
34	2.732	.3660	.0173	.0473	57.730	21.132	13.702	289.544	**34**
35	2.814	.3554	.0165	.0465	60.462	21.487	14.037	301.627	**35**
40	3.262	.3066	.0133	.0433	75.401	23.115	15.650	361.750	**40**
45	3.782	.2644	.0108	.0408	92.720	24.519	17.156	420.632	**45**
50	4.384	.2281	.00887	.0389	112.797	25.730	18.558	477.480	**50**
55	5.082	.1968	.00735	.0373	136.072	26.774	19.860	531.741	**55**
60	5.892	.1697	.00613	.0361	163.053	27.676	21.067	583.052	**60**
65	6.830	.1464	.00515	.0351	194.333	28.453	22.184	631.201	**65**
70	7.918	.1263	.00434	.0343	230.594	29.123	23.215	676.087	**70**
75	9.179	.1089	.00367	.0337	272.631	29.702	24.163	717.698	**75**
80	10.641	.0940	.00311	.0331	321.363	30.201	25.035	756.086	**80**
85	12.336	.0811	.00265	.0326	377.857	30.631	25.835	791.353	**85**
90	14.300	.0699	.00226	.0323	443.349	31.002	26.567	823.630	**90**
95	16.578	.0603	.00193	.0319	519.272	31.323	27.235	853.074	**95**
100	19.219	.0520	.00165	.0316	607.287	31.599	27.844	879.854	**100**

Compound Interest Factors

3½%

	Single Payment		Uniform Payment Series				Arithmetic Gradient		
	Compound Amount Factor	Present Worth Factor	Sinking Fund Factor	Capital Recovery Factor	Compound Amount Factor	Present Worth Factor	Gradient Uniform Series	Gradient Present Worth	
n	Find *F* Given *P* *F/P*	Find *P* Given *F* *P/F*	Find *A* Given *F* *A/F*	Find *A* Given *P* *A/P*	Find *F* Given *A* *F/A*	Find *P* Given *A* *P/A*	Find *A* Given *G* *A/G*	Find *P* Given *G* *P/G*	*n*
1	1.035	.9662	1.0000	1.0350	1.000	0.966	0	0	**1**
2	1.071	.9335	.4914	.5264	2.035	1.900	0.491	0.933	**2**
3	1.109	.9019	.3219	.3569	3.106	2.802	0.977	2.737	**3**
4	1.148	.8714	.2373	.2723	4.215	3.673	1.457	5.352	**4**
5	1.188	.8420	.1865	.2215	5.362	4.515	1.931	8.719	**5**
6	1.229	.8135	.1527	.1877	6.550	5.329	2.400	12.787	**6**
7	1.272	.7860	.1285	.1635	7.779	6.115	2.862	17.503	**7**
8	1.317	.7594	.1105	.1455	9.052	6.874	3.320	22.819	**8**
9	1.363	.7337	.0964	.1314	10.368	7.608	3.771	28.688	**9**
10	1.411	.7089	.0852	.1202	11.731	8.317	4.217	35.069	**10**
11	1.460	.6849	.0761	.1111	13.142	9.002	4.657	41.918	**11**
12	1.511	.6618	.0685	.1035	14.602	9.663	5.091	49.198	**12**
13	1.564	.6394	.0621	.0971	16.113	10.303	5.520	56.871	**13**
14	1.619	.6178	.0566	.0916	17.677	10.921	5.943	64.902	**14**
15	1.675	.5969	.0518	.0868	19.296	11.517	6.361	73.258	**15**
16	1.734	.5767	.0477	.0827	20.971	12.094	6.773	81.909	**16**
17	1.795	.5572	.0440	.0790	22.705	12.651	7.179	90.824	**17**
18	1.857	.5384	.0408	.0758	24.500	13.190	7.580	99.976	**18**
19	1.922	.5202	.0379	.0729	26.357	13.710	7.975	109.339	**19**
20	1.990	.5026	.0354	.0704	28.280	14.212	8.365	118.888	**20**
21	2.059	.4856	.0330	.0680	30.269	14.698	8.749	128.599	**21**
22	2.132	.4692	.0309	.0659	32.329	15.167	9.128	138.451	**22**
23	2.206	.4533	.0290	.0640	34.460	15.620	9.502	148.423	**23**
24	2.283	.4380	.0273	.0623	36.666	16.058	9.870	158.496	**24**
25	2.363	.4231	.0257	.0607	38.950	16.482	10.233	168.652	**25**
26	2.446	.4088	.0242	.0592	41.313	16.890	10.590	178.873	**26**
27	2.532	.3950	.0229	.0579	43.759	17.285	10.942	189.143	**27**
28	2.620	.3817	.0216	.0566	46.291	17.667	11.289	199.448	**28**
29	2.712	.3687	.0204	.0554	48.911	18.036	11.631	209.773	**29**
30	2.807	.3563	.0194	.0544	51.623	18.392	11.967	220.105	**30**
31	2.905	.3442	.0184	.0534	54.429	18.736	12.299	230.432	**31**
32	3.007	.3326	.0174	.0524	57.334	19.069	12.625	240.742	**32**
33	3.112	.3213	.0166	.0516	60.341	19.390	12.946	251.025	**33**
34	3.221	.3105	.0158	.0508	63.453	19.701	13.262	261.271	**34**
35	3.334	.3000	.0150	.0500	66.674	20.001	13.573	271.470	**35**
40	3.959	.2526	.0118	.0468	84.550	21.355	15.055	321.490	**40**
45	4.702	.2127	.00945	.0445	105.781	22.495	16.417	369.307	**45**
50	5.585	.1791	.00763	.0426	130.998	23.456	17.666	414.369	**50**
55	6.633	.1508	.00621	.0412	160.946	24.264	18.808	456.352	**55**
60	7.878	.1269	.00509	.0401	196.516	24.945	19.848	495.104	**60**
65	9.357	.1069	.00419	.0392	238.762	25.518	20.793	530.598	**65**
70	11.113	.0900	.00346	.0385	288.937	26.000	21.650	562.895	**70**
75	13.199	.0758	.00287	.0379	348.529	26.407	22.423	592.121	**75**
80	15.676	.0638	.00238	.0374	419.305	26.749	23.120	618.438	**80**
85	18.618	.0537	.00199	.0370	503.365	27.037	23.747	642.036	**85**
90	22.112	.0452	.00166	.0367	603.202	27.279	24.308	663.118	**90**
95	26.262	.0381	.00139	.0364	721.778	27.483	24.811	681.890	**95**
100	31.191	.0321	.00116	.0362	862.608	27.655	25.259	698.554	**100**

n	Single Payment		Uniform Payment Series				Arithmetic Gradient		n
	Compound Amount Factor	Present Worth Factor	Sinking Fund Factor	Capital Recovery Factor	Compound Amount Factor	Present Worth Factor	Gradient Uniform Series	Gradient Present Worth	
	Find F Given P F/P	Find P Given F P/F	Find A Given F A/F	Find A Given P A/P	Find F Given A F/A	Find P Given A P/A	Find A Given G A/G	Find P Given G P/G	
1	1.040	.9615	1.0000	1.0400	1.000	0.962	0	0	**1**
2	1.082	.9246	.4902	.5302	2.040	1.886	0.490	0.925	**2**
3	1.125	.8890	.3203	.3603	3.122	2.775	0.974	2.702	**3**
4	1.170	.8548	.2355	.2755	4.246	3.630	1.451	5.267	**4**
5	1.217	.8219	.1846	.2246	5.416	4.452	1.922	8.555	**5**
6	1.265	.7903	.1508	.1908	6.633	5.242	2.386	12.506	**6**
7	1.316	.7599	.1266	.1666	7.898	6.002	2.843	17.066	**7**
8	1.369	.7307	.1085	.1485	9.214	6.733	3.294	22.180	**8**
9	1.423	.7026	.0945	.1345	10.583	7.435	3.739	27.801	**9**
10	1.480	.6756	.0833	.1233	12.006	8.111	4.177	33.881	**10**
11	1.539	.6496	.0741	.1141	13.486	8.760	4.609	40.377	**11**
12	1.601	.6246	.0666	.1066	15.026	9.385	5.034	47.248	**12**
13	1.665	.6006	.0601	.1001	16.627	9.986	5.453	54.454	**13**
14	1.732	.5775	.0547	.0947	18.292	10.563	5.866	61.962	**14**
15	1.801	.5553	.0499	.0899	20.024	11.118	6.272	69.735	**15**
16	1.873	.5339	.0458	.0858	21.825	11.652	6.672	77.744	**16**
17	1.948	.5134	.0422	.0822	23.697	12.166	7.066	85.958	**17**
18	2.026	.4936	.0390	.0790	25.645	12.659	7.453	94.350	**18**
19	2.107	.4746	.0361	.0761	27.671	13.134	7.834	102.893	**19**
20	2.191	.4564	.0336	.0736	29.778	13.590	8.209	111.564	**20**
21	2.279	.4388	.0313	.0713	31.969	14.029	8.578	120.341	**21**
22	2.370	.4220	.0292	.0692	34.248	14.451	8.941	129.202	**22**
23	2.465	.4057	.0273	.0673	36.618	14.857	9.297	138.128	**23**
24	2.563	.3901	.0256	.0656	39.083	15.247	9.648	147.101	**24**
25	2.666	.3751	.0240	.0640	41.646	15.622	9.993	156.104	**25**
26	2.772	.3607	.0226	.0626	44.312	15.983	10.331	165.121	**26**
27	2.883	.3468	.0212	.0612	47.084	16.330	10.664	174.138	**27**
28	2.999	.3335	.0200	.0600	49.968	16.663	10.991	183.142	**28**
29	3.119	.3207	.0189	.0589	52.966	16.984	11.312	192.120	**29**
30	3.243	.3083	.0178	.0578	56.085	17.292	11.627	201.062	**30**
31	3.373	.2965	.0169	.0569	59.328	17.588	11.937	209.955	**31**
32	3.508	.2851	.0159	.0559	62.701	17.874	12.241	218.792	**32**
33	3.648	.2741	.0151	.0551	66.209	18.148	12.540	227.563	**33**
34	3.794	.2636	.0143	.0543	69.858	18.411	12.832	236.260	**34**
35	3.946	.2534	.0136	.0536	73.652	18.665	13.120	244.876	**35**
40	4.801	.2083	.0105	.0505	95.025	19.793	14.476	286.530	**40**
45	5.841	.1712	.00826	.0483	121.029	20.720	15.705	325.402	**45**
50	7.107	.1407	.00655	.0466	152.667	21.482	16.812	361.163	**50**
55	8.646	.1157	.00523	.0452	191.159	22.109	17.807	393.689	**55**
60	10.520	.0951	.00420	.0442	237.990	22.623	18.697	422.996	**60**
65	12.799	.0781	.00339	.0434	294.968	23.047	19.491	449.201	**65**
70	15.572	.0642	.00275	.0427	364.290	23.395	20.196	472.479	**70**
75	18.945	.0528	.00223	.0422	448.630	23.680	20.821	493.041	**75**
80	23.050	.0434	.00181	.0418	551.244	23.915	21.372	511.116	**80**
85	28.044	.0357	.00148	.0415	676.089	24.109	21.857	526.938	**85**
90	34.119	.0293	.00121	.0412	827.981	24.267	22.283	540.737	**90**
95	41.511	.0241	.00099	.0410	1 012.8	24.398	22.655	552.730	**95**
100	50.505	.0198	.00081	.0408	1 237.6	24.505	22.980	563.125	**100**

	Single Payment		Uniform Payment Series				Arithmetic Gradient		
	Compound Amount Factor	Present Worth Factor	Sinking Fund Factor	Capital Recovery Factor	Compound Amount Factor	Present Worth Factor	Gradient Uniform Series	Gradient Present Worth	
n	Find *F* Given *P* *F/P*	Find *P* Given *F* *P/F*	Find *A* Given *F* *A/F*	Find *A* Given *P* *A/P*	Find *F* Given *A* *F/A*	Find *P* Given *A* *P/A*	Find *A* Given *G* *A/G*	Find *P* Given *G* *P/G*	*n*
1	1.045	.9569	1.0000	1.0450	1.000	0.957	0	0	**1**
2	1.092	.9157	.4890	.5340	2.045	1.873	0.489	0.916	**2**
3	1.141	.8763	.3188	.3638	3.137	2.749	0.971	2.668	**3**
4	1.193	.8386	.2337	.2787	4.278	3.588	1.445	5.184	**4**
5	1.246	.8025	.1828	.2278	5.471	4.390	1.912	8.394	**5**
6	1.302	.7679	.1489	.1939	6.717	5.158	2.372	12.233	**6**
7	1.361	.7348	.1247	.1697	8.019	5.893	2.824	16.642	**7**
8	1.422	.7032	.1066	.1516	9.380	6.596	3.269	21.564	**8**
9	1.486	.6729	.0926	.1376	10.802	7.269	3.707	26.948	**9**
10	1.553	.6439	.0814	.1264	12.288	7.913	4.138	32.743	**10**
11	1.623	.6162	.0722	.1172	13.841	8.529	4.562	38.905	**11**
12	1.696	.5897	.0647	.1097	15.464	9.119	4.978	45.391	**12**
13	1.772	.5643	.0583	.1033	17.160	9.683	5.387	52.163	**13**
14	1.852	.5400	.0528	.0978	18.932	10.223	5.789	59.182	**14**
15	1.935	.5167	.0481	.0931	20.784	10.740	6.184	66.416	**15**
16	2.022	.4945	.0440	.0890	22.719	11.234	6.572	73.833	**16**
17	2.113	.4732	.0404	.0854	24.742	11.707	6.953	81.404	**17**
18	2.208	.4528	.0372	.0822	26.855	12.160	7.327	89.102	**18**
19	2.308	.4333	.0344	.0794	29.064	12.593	7.695	96.901	**19**
20	2.412	.4146	.0319	.0769	31.371	13.008	8.055	104.779	**20**
21	2.520	.3968	.0296	.0746	33.783	13.405	8.409	112.715	**21**
22	2.634	.3797	.0275	.0725	36.303	13.784	8.755	120.689	**22**
23	2.752	.3634	.0257	.0707	38.937	14.148	9.096	128.682	**23**
24	2.876	.3477	.0240	.0690	41.689	14.495	9.429	136.680	**24**
25	3.005	.3327	.0224	.0674	44.565	14.828	9.756	144.665	**25**
26	3.141	.3184	.0210	.0660	47.571	15.147	10.077	152.625	**26**
27	3.282	.3047	.0197	.0647	50.711	15.451	10.391	160.547	**27**
28	3.430	.2916	.0185	.0635	53.993	15.743	10.698	168.420	**28**
29	3.584	.2790	.0174	.0624	57.423	16.022	10.999	176.232	**29**
30	3.745	.2670	.0164	.0614	61.007	16.289	11.295	183.975	**30**
31	3.914	.2555	.0154	.0604	64.752	16.544	11.583	191.640	**31**
32	4.090	.2445	.0146	.0596	68.666	16.789	11.866	199.220	**32**
33	4.274	.2340	.0137	.0587	72.756	17.023	12.143	206.707	**33**
34	4.466	.2239	.0130	.0580	77.030	17.247	12.414	214.095	**34**
35	4.667	.2143	.0123	.0573	81.497	17.461	12.679	221.380	**35**
40	5.816	.1719	.00934	.0543	107.030	18.402	13.917	256.098	**40**
45	7.248	.1380	.00720	.0522	138.850	19.156	15.020	287.732	**45**
50	9.033	.1107	.00560	.0506	178.503	19.762	15.998	316.145	**50**
55	11.256	.0888	.00439	.0494	227.918	20.248	16.860	341.375	**55**
60	14.027	.0713	.00345	.0485	289.497	20.638	17.617	363.571	**60**
65	17.481	.0572	.00273	.0477	366.237	20.951	18.278	382.946	**65**
70	21.784	.0459	.00217	.0472	461.869	21.202	18.854	399.750	**70**
75	27.147	.0368	.00172	.0467	581.043	21.404	19.354	414.242	**75**
80	33.830	.0296	.00137	.0464	729.556	21.565	19.785	426.680	**80**
85	42.158	.0237	.00109	.0461	914.630	21.695	20.157	437.309	**85**
90	52.537	.0190	.00087	.0459	1 145.3	21.799	20.476	446.359	**90**
95	65.471	.0153	.00070	.0457	1 432.7	21.883	20.749	454.039	**95**
100	81.588	.0123	.00056	.0456	1 790.9	21.950	20.981	460.537	**100**

	Single Payment		Uniform Payment Series				Arithmetic Gradient		
	Compound Amount Factor	Present Worth Factor	Sinking Fund Factor	Capital Recovery Factor	Compound Amount Factor	Present Worth Factor	Gradient Uniform Series	Gradient Present Worth	
n	Find *F* Given *P* *F/P*	Find *P* Given *F* *P/F*	Find *A* Given *F* *A/F*	Find *A* Given *P* *A/P*	Find *F* Given *A* *F/A*	Find *P* Given *A* *P/A*	Find *A* Given *G* *A/G*	Find *P* Given *G* *P/G*	n
1	1.050	.9524	1.0000	1.0500	1.000	0.952	0	0	**1**
2	1.102	.9070	.4878	.5378	2.050	1.859	0.488	0.907	**2**
3	1.158	.8638	.3172	.3672	3.152	2.723	0.967	2.635	**3**
4	1.216	.8227	.2320	.2820	4.310	3.546	1.439	5.103	**4**
5	1.276	.7835	.1810	.2310	5.526	4.329	1.902	8.237	**5**
6	1.340	.7462	.1470	.1970	6.802	5.076	2.358	11.968	**6**
7	1.407	.7107	.1228	.1728	8.142	5.786	2.805	16.232	**7**
8	1.477	.6768	.1047	.1547	9.549	6.463	3.244	20.970	**8**
9	1.551	.6446	.0907	.1407	11.027	7.108	3.676	26.127	**9**
10	1.629	.6139	.0795	.1295	12.578	7.722	4.099	31.652	**10**
11	1.710	.5847	.0704	.1204	14.207	8.306	4.514	37.499	**11**
12	1.796	.5568	.0628	.1128	15.917	8.863	4.922	43.624	**12**
13	1.886	.5303	.0565	.1065	17.713	9.394	5.321	49.988	**13**
14	1.980	.5051	.0510	.1010	19.599	9.899	5.713	56.553	**14**
15	2.079	.4810	.0463	.0963	21.579	10.380	6.097	63.288	**15**
16	2.183	.4581	.0423	.0923	23.657	10.838	6.474	70.159	**16**
17	2.292	.4363	.0387	.0887	25.840	11.274	6.842	77.140	**17**
18	2.407	.4155	.0355	.0855	28.132	11.690	7.203	84.204	**18**
19	2.527	.3957	.0327	.0827	30.539	12.085	7.557	91.327	**19**
20	2.653	.3769	.0302	.0802	33.066	12.462	7.903	98.488	**20**
21	2.786	.3589	.0280	.0780	35.719	12.821	8.242	105.667	**21**
22	2.925	.3419	.0260	.0760	38.505	13.163	8.573	112.846	**22**
23	3.072	.3256	.0241	.0741	41.430	13.489	8.897	120.008	**23**
24	3.225	.3101	.0225	.0725	44.502	13.799	9.214	127.140	**24**
25	3.386	.2953	.0210	.0710	47.727	14.094	9.524	134.227	**25**
26	3.556	.2812	.0196	.0696	51.113	14.375	9.827	141.258	**26**
27	3.733	.2678	.0183	.0683	54.669	14.643	10.122	148.222	**27**
28	3.920	.2551	.0171	.0671	58.402	14.898	10.411	155.110	**28**
29	4.116	.2429	.0160	.0660	62.323	15.141	10.694	161.912	**29**
30	4.322	.2314	.0151	.0651	66.439	15.372	10.969	168.622	**30**
31	4.538	.2204	.0141	.0641	70.761	15.593	11.238	175.233	**31**
32	4.765	.2099	.0133	.0633	75.299	15.803	11.501	181.739	**32**
33	5.003	.1999	.0125	.0625	80.063	16.003	11.757	188.135	**33**
34	5.253	.1904	.0118	.0618	85.067	16.193	12.006	194.416	**34**
35	5.516	.1813	.0111	.0611	90.320	16.374	12.250	200.580	**35**
40	7.040	.1420	.00828	.0583	120.799	17.159	13.377	229.545	**40**
45	8.985	.1113	.00626	.0563	159.699	17.774	14.364	255.314	**45**
50	11.467	.0872	.00478	.0548	209.347	18.256	15.223	277.914	**50**
55	14.636	.0683	.00367	.0537	272.711	18.633	15.966	297.510	**55**
60	18.679	.0535	.00283	.0528	353.582	18.929	16.606	314.343	**60**
65	23.840	.0419	.00219	.0522	456.795	19.161	17.154	328.691	**65**
70	30.426	.0329	.00170	.0517	588.525	19.343	17.621	340.841	**70**
75	38.832	.0258	.00132	.0513	756.649	19.485	18.018	351.072	**75**
80	49.561	.0202	.00103	.0510	971.222	19.596	18.353	359.646	**80**
85	63.254	.0158	.00080	.0508	1 245.1	19.684	18.635	366.800	**85**
90	80.730	.0124	.00063	.0506	1 594.6	19.752	18.871	372.749	**90**
95	103.034	.00971	.00049	.0505	2 040.7	19.806	19.069	377.677	**95**
100	131.500	.00760	.00038	.0504	2 610.0	19.848	19.234	381.749	**100**

	Single Payment		Uniform Payment Series				Arithmetic Gradient		
	Compound Amount Factor	Present Worth Factor	Sinking Fund Factor	Capital Recovery Factor	Compound Amount Factor	Present Worth Factor	Gradient Uniform Series	Gradient Present Worth	
n	Find F Given P F/P	Find P Given F P/F	Find A Given F A/F	Find A Given P A/P	Find F Given A F/A	Find P Given A P/A	Find A Given G A/G	Find P Given G P/G	n
1	1.060	.9434	1.0000	1.0600	1.000	0.943	0	0	**1**
2	1.124	.8900	.4854	.5454	2.060	1.833	0.485	0.890	**2**
3	1.191	.8396	.3141	.3741	3.184	2.673	0.961	2.569	**3**
4	1.262	.7921	.2286	.2886	4.375	3.465	1.427	4.945	**4**
5	1.338	.7473	.1774	.2374	5.637	4.212	1.884	7.934	**5**
6	1.419	.7050	.1434	.2034	6.975	4.917	2.330	11.459	**6**
7	1.504	.6651	.1191	.1791	8.394	5.582	2.768	15.450	**7**
8	1.594	.6274	.1010	.1610	9.897	6.210	3.195	19.841	**8**
9	1.689	.5919	.0870	.1470	11.491	6.802	3.613	24.577	**9**
10	1.791	.5584	.0759	.1359	13.181	7.360	4.022	29.602	**10**
11	1.898	.5268	.0668	.1268	14.972	7.887	4.421	34.870	**11**
12	2.012	.4970	.0593	.1193	16.870	8.384	4.811	40.337	**12**
13	2.133	.4688	.0530	.1130	18.882	8.853	5.192	45.963	**13**
14	2.261	.4423	.0476	.1076	21.015	9.295	5.564	51.713	**14**
15	2.397	.4173	.0430	.1030	23.276	9.712	5.926	57.554	**15**
16	2.540	.3936	.0390	.0990	25.672	10.106	6.279	63.459	**16**
17	2.693	.3714	.0354	.0954	28.213	10.477	6.624	69.401	**17**
18	2.854	.3503	.0324	.0924	30.906	10.828	6.960	75.357	**18**
19	3.026	.3305	.0296	.0896	33.760	11.158	7.287	81.306	**19**
20	3.207	.3118	.0272	.0872	36.786	11.470	7.605	87.230	**20**
21	3.400	.2942	.0250	.0850	39.993	11.764	7.915	93.113	**21**
22	3.604	.2775	.0230	.0830	43.392	12.042	8.217	98.941	**22**
23	3.820	.2618	.0213	.0813	46.996	12.303	8.510	104.700	**23**
24	4.049	.2470	.0197	.0797	50.815	12.550	8.795	110.381	**24**
25	4.292	.2330	.0182	.0782	54.864	12.783	9.072	115.973	**25**
26	4.549	.2198	.0169	.0769	59.156	13.003	9.341	121.468	**26**
27	4.822	.2074	.0157	.0757	63.706	13.211	9.603	126.860	**27**
28	5.112	.1956	.0146	.0746	68.528	13.406	9.857	132.142	**28**
29	5.418	.1846	.0136	.0736	73.640	13.591	10.103	137.309	**29**
30	5.743	.1741	.0126	.0726	79.058	13.765	10.342	142.359	**30**
31	6.088	.1643	.0118	.0718	84.801	13.929	10.574	147.286	**31**
32	6.453	.1550	.0110	.0710	90.890	14.084	10.799	152.090	**32**
33	6.841	.1462	.0103	.0703	97.343	14.230	11.017	156.768	**33**
34	7.251	.1379	.00960	.0696	104.184	14.368	11.228	161.319	**34**
35	7.686	.1301	.00897	.0690	111.435	14.498	11.432	165.743	**35**
40	10.286	.0972	.00646	.0665	154.762	15.046	12.359	185.957	**40**
45	13.765	.0727	.00470	.0647	212.743	15.456	13.141	203.109	**45**
50	18.420	.0543	.00344	.0634	290.335	15.762	13.796	217.457	**50**
55	24.650	.0406	.00254	.0625	394.171	15.991	14.341	229.322	**55**
60	32.988	.0303	.00188	.0619	533.126	16.161	14.791	239.043	**60**
65	44.145	.0227	.00139	.0614	719.080	16.289	15.160	246.945	**65**
70	59.076	.0169	.00103	.0610	967.928	16.385	15.461	253.327	**70**
75	79.057	.0126	.00077	.0608	1 300.9	16.456	15.706	258.453	**75**
80	105.796	.00945	.00057	.0606	1 746.6	16.509	15.903	262.549	**80**
85	141.578	.00706	.00043	.0604	2 343.0	16.549	16.062	265.810	**85**
90	189.464	.00528	.00032	.0603	3 141.1	16.579	16.189	268.395	**90**
95	253.545	.00394	.00024	.0602	4 209.1	16.601	16.290	270.437	**95**
100	339.300	.00295	.00018	.0602	5 638.3	16.618	16.371	272.047	**100**

7% Compound Interest Factors 7%

	Single Payment		Uniform Payment Series				Arithmetic Gradient		
	Compound Amount Factor	Present Worth Factor	Sinking Fund Factor	Capital Recovery Factor	Compound Amount Factor	Present Worth Factor	Gradient Uniform Series	Gradient Present Worth	
n	Find *F* Given *P* *F/P*	Find *P* Given *F* *P/F*	Find *A* Given *F* *A/F*	Find *A* Given *P* *A/P*	Find *F* Given *A* *F/A*	Find *P* Given *A* *P/A*	Find *A* Given *G* *A/G*	Find *P* Given *G* *P/G*	**n**
1	1.070	.9346	1.0000	1.0700	1.000	0.935	0	0	**1**
2	1.145	.8734	.4831	.5531	2.070	1.808	0.483	0.873	**2**
3	1.225	.8163	.3111	.3811	3.215	2.624	0.955	2.506	**3**
4	1.311	.7629	.2252	.2952	4.440	3.387	1.416	4.795	**4**
5	1.403	.7130	.1739	.2439	5.751	4.100	1.865	7.647	**5**
6	1.501	.6663	.1398	.2098	7.153	4.767	2.303	10.978	**6**
7	1.606	.6227	.1156	.1856	8.654	5.389	2.730	14.715	**7**
8	1.718	.5820	.0975	.1675	10.260	5.971	3.147	18.789	**8**
9	1.838	.5439	.0835	.1535	11.978	6.515	3.552	23.140	**9**
10	1.967	.5083	.0724	.1424	13.816	7.024	3.946	27.716	**10**
11	2.105	.4751	.0634	.1334	15.784	7.499	4.330	32.467	**11**
12	2.252	.4440	.0559	.1259	17.888	7.943	4.703	37.351	**12**
13	2.410	.4150	.0497	.1197	20.141	8.358	5.065	42.330	**13**
14	2.579	.3878	.0443	.1143	22.551	8.745	5.417	47.372	**14**
15	2.759	.3624	.0398	.1098	25.129	9.108	5.758	52.446	**15**
16	2.952	.3387	.0359	.1059	27.888	9.447	6.090	57.527	**16**
17	3.159	.3166	.0324	.1024	30.840	9.763	6.411	62.592	**17**
18	3.380	.2959	.0294	.0994	33.999	10.059	6.722	67.622	**18**
19	3.617	.2765	.0268	.0968	37.379	10.336	7.024	72.599	**19**
20	3.870	.2584	.0244	.0944	40.996	10.594	7.316	77.509	**20**
21	4.141	.2415	.0223	.0923	44.865	10.836	7.599	82.339	**21**
22	4.430	.2257	.0204	.0904	49.006	11.061	7.872	87.079	**22**
23	4.741	.2109	.0187	.0887	53.436	11.272	8.137	91.720	**23**
24	5.072	.1971	.0172	.0872	58.177	11.469	8.392	96.255	**24**
25	5.427	.1842	.0158	.0858	63.249	11.654	8.639	100.677	**25**
26	5.807	.1722	.0146	.0846	68.677	11.826	8.877	104.981	**26**
27	6.214	.1609	.0134	.0834	74.484	11.987	9.107	109.166	**27**
28	6.649	.1504	.0124	.0824	80.698	12.137	9.329	113.227	**28**
29	7.114	.1406	.0114	.0814	87.347	12.278	9.543	117.162	**29**
30	7.612	.1314	.0106	.0806	94.461	12.409	9.749	120.972	**30**
31	8.145	.1228	.00980	.0798	102.073	12.532	9.947	124.655	**31**
32	8.715	.1147	.00907	.0791	110.218	12.647	10.138	128.212	**32**
33	9.325	.1072	.00841	.0784	118.934	12.754	10.322	131.644	**33**
34	9.978	.1002	.00780	.0778	128.259	12.854	10.499	134.951	**34**
35	10.677	.0937	.00723	.0772	138.237	12.948	10.669	138.135	**35**
40	14.974	.0668	.00501	.0750	199.636	13.332	11.423	152.293	**40**
45	21.002	.0476	.00350	.0735	285.750	13.606	12.036	163.756	**45**
50	29.457	.0339	.00246	.0725	406.530	13.801	12.529	172.905	**50**
55	41.315	.0242	.00174	.0717	575.930	13.940	12.921	180.124	**55**
60	57.947	.0173	.00123	.0712	813.523	14.039	13.232	185.768	**60**
65	81.273	.0123	.00087	.0709	1 146.8	14.110	13.476	190.145	**65**
70	113.990	.00877	.00062	.0706	1 614.1	14.160	13.666	193.519	**70**
75	159.877	.00625	.00044	.0704	2 269.7	14.196	13.814	196.104	**75**
80	224.235	.00446	.00031	.0703	3 189.1	14.222	13.927	198.075	**80**
85	314.502	.00318	.00022	.0702	4 478.6	14.240	14.015	199.572	**85**
90	441.105	.00227	.00016	.0702	6 287.2	14.253	14.081	200.704	**90**
95	618.673	.00162	.00011	.0701	8 823.9	14.263	14.132	201.558	**95**
100	867.720	.00115	.00008	.0701	12 381.7	14.269	14.170	202.200	**100**

	Single Payment		Uniform Payment Series				Arithmetic Gradient		
	Compound Amount Factor	Present Worth Factor	Sinking Fund Factor	Capital Recovery Factor	Compound Amount Factor	Present Worth Factor	Gradient Uniform Series	Gradient Present Worth	
n	Find *F* Given *P* *F/P*	Find *P* Given *F* *P/F*	Find *A* Given *F* *A/F*	Find *A* Given *P* *A/P*	Find *F* Given *A* *F/A*	Find *P* Given *A* *P/A*	Find *A* Given *G* *A/G*	Find *P* Given *G* *P/G*	*n*
1	1.080	.9259	1.0000	1.0800	1.000	0.926	0	0	**1**
2	1.166	.8573	.4808	.5608	2.080	1.783	0.481	0.857	**2**
3	1.260	.7938	.3080	.3880	3.246	2.577	0.949	2.445	**3**
4	1.360	.7350	.2219	.3019	4.506	3.312	1.404	4.650	**4**
5	1.469	.6806	.1705	.2505	5.867	3.993	1.846	7.372	**5**
6	1.587	.6302	.1363	.2163	7.336	4.623	2.276	10.523	**6**
7	1.714	.5835	.1121	.1921	8.923	5.206	2.694	14.024	**7**
8	1.851	.5403	.0940	.1740	10.637	5.747	3.099	17.806	**8**
9	1.999	.5002	.0801	.1601	12.488	6.247	3.491	21.808	**9**
10	2.159	.4632	.0690	.1490	14.487	6.710	3.871	25.977	**10**
11	2.332	.4289	.0601	.1401	16.645	7.139	4.240	30.266	**11**
12	2.518	.3971	.0527	.1327	18.977	7.536	4.596	34.634	**12**
13	2.720	.3677	.0465	.1265	21.495	7.904	4.940	39.046	**13**
14	2.937	.3405	.0413	.1213	24.215	8.244	5.273	43.472	**14**
15	3.172	.3152	.0368	.1168	27.152	8.559	5.594	47.886	**15**
16	3.426	.2919	.0330	.1130	30.324	8.851	5.905	52.264	**16**
17	3.700	.2703	.0296	.1096	33.750	9.122	6.204	56.588	**17**
18	3.996	.2502	.0267	.1067	37.450	9.372	6.492	60.843	**18**
19	4.316	.2317	.0241	.1041	41.446	9.604	6.770	65.013	**19**
20	4.661	.2145	.0219	.1019	45.762	9.818	7.037	69.090	**20**
21	5.034	.1987	.0198	.0998	50.423	10.017	7.294	73.063	**21**
22	5.437	.1839	.0180	.0980	55.457	10.201	7.541	76.926	**22**
23	5.871	.1703	.0164	.0964	60.893	10.371	7.779	80.673	**23**
24	6.341	.1577	.0150	.0950	66.765	10.529	8.007	84.300	**24**
25	6.848	.1460	.0137	.0937	73.106	10.675	8.225	87.804	**25**
26	7.396	.1352	.0125	.0925	79.954	10.810	8.435	91.184	**26**
27	7.988	.1252	.0114	.0914	87.351	10.935	8.636	94.439	**27**
28	8.627	.1159	.0105	.0905	95.339	11.051	8.829	97.569	**28**
29	9.317	.1073	.00962	.0896	103.966	11.158	9.013	100.574	**29**
30	10.063	.0994	.00883	.0888	113.283	11.258	9.190	103.456	**30**
31	10.868	.0920	.00811	.0881	123.346	11.350	9.358	106.216	**31**
32	11.737	.0852	.00745	.0875	134.214	11.435	9.520	108.858	**32**
33	12.676	.0789	.00685	.0869	145.951	11.514	9.674	111.382	**33**
34	13.690	.0730	.00630	.0863	158.627	11.587	9.821	113.792	**34**
35	14.785	.0676	.00580	.0858	172.317	11.655	9.961	116.092	**35**
40	21.725	.0460	.00386	.0839	259.057	11.925	10.570	126.042	**40**
45	31.920	.0313	.00259	.0826	386.506	12.108	11.045	133.733	**45**
50	46.902	.0213	.00174	.0817	573.771	12.233	11.411	139.593	**50**
55	68.914	.0145	.00118	.0812	848.925	12.319	11.690	144.006	**55**
60	101.257	.00988	.00080	.0808	1 253.2	12.377	11.902	147.300	**60**
65	148.780	.00672	.00054	.0805	1 847.3	12.416	12.060	149.739	**65**
70	218.607	.00457	.00037	.0804	2 720.1	12.443	12.178	151.533	**70**
75	321.205	.00311	.00025	.0802	4 002.6	12.461	12.266	152.845	**75**
80	471.956	.00212	.00017	.0802	5 887.0	12.474	12.330	153.800	**80**
85	693.458	.00144	.00012	.0801	8 655.7	12.482	12.377	154.492	**85**
90	1 018.9	.00098	.00008	.0801	12 724.0	12.488	12.412	154.993	**90**
95	1 497.1	.00067	.00005	.0801	18 701.6	12.492	12.437	155.352	**95**
100	2 199.8	.00045	.00004	.0800	27 484.6	12.494	12.455	155.611	**100**

	Single Payment		Uniform Payment Series				Arithmetic Gradient		
	Compound Amount Factor	Present Worth Factor	Sinking Fund Factor	Capital Recovery Factor	Compound Amount Factor	Present Worth Factor	Gradient Uniform Series	Gradient Present Worth	
n	Find *F* Given *P* *F/P*	Find *P* Given *F* *P/F*	Find *A* Given *F* *A/F*	Find *A* Given *P* *A/P*	Find *F* Given *A* *F/A*	Find *P* Given *A* *P/A*	Find *A* Given *G* *A/G*	Find *P* Given *G* *P/G*	*n*
1	1.090	.9174	1.0000	1.0900	1.000	0.917	0	0	**1**
2	1.188	.8417	.4785	.5685	2.090	1.759	0.478	0.842	**2**
3	1.295	.7722	.3051	.3951	3.278	2.531	0.943	2.386	**3**
4	1.412	.7084	.2187	.3087	4.573	3.240	1.393	4.511	**4**
5	1.539	.6499	.1671	.2571	5.985	3.890	1.828	7.111	**5**
6	1.677	.5963	.1329	.2229	7.523	4.486	2.250	10.092	**6**
7	1.828	.5470	.1087	.1987	9.200	5.033	2.657	13.375	**7**
8	1.993	.5019	.0907	.1807	11.028	5.535	3.051	16.888	**8**
9	2.172	.4604	.0768	.1668	13.021	5.995	3.431	20.571	**9**
10	2.367	.4224	.0658	.1558	15.193	6.418	3.798	24.373	**10**
11	2.580	.3875	.0569	.1469	17.560	6.805	4.151	28.248	**11**
12	2.813	.3555	.0497	.1397	20.141	7.161	4.491	32.159	**12**
13	3.066	.3262	.0436	.1336	22.953	7.487	4.818	36.073	**13**
14	3.342	.2992	.0384	.1284	26.019	7.786	5.133	39.963	**14**
15	3.642	.2745	.0341	.1241	29.361	8.061	5.435	43.807	**15**
16	3.970	.2519	.0303	.1203	33.003	8.313	5.724	47.585	**16**
17	4.328	.2311	.0270	.1170	36.974	8.544	6.002	51.282	**17**
18	4.717	.2120	.0242	.1142	41.301	8.756	6.269	54.886	**18**
19	5.142	.1945	.0217	.1117	46.019	8.950	6.524	58.387	**19**
20	5.604	.1784	.0195	.1095	51.160	9.129	6.767	61.777	**20**
21	6.109	.1637	.0176	.1076	56.765	9.292	7.001	65.051	**21**
22	6.659	.1502	.0159	.1059	62.873	9.442	7.223	68.205	**22**
23	7.258	.1378	.0144	.1044	69.532	9.580	7.436	71.236	**23**
24	7.911	.1264	.0130	.1030	76.790	9.707	7.638	74.143	**24**
25	8.623	.1160	.0118	.1018	84.701	9.823	7.832	76.927	**25**
26	9.399	.1064	.0107	.1007	93.324	9.929	8.016	79.586	**26**
27	10.245	.0976	.00973	.0997	102.723	10.027	8.191	82.124	**27**
28	11.167	.0895	.00885	.0989	112.968	10.116	8.357	84.542	**28**
29	12.172	.0822	.00806	.0981	124.136	10.198	8.515	86.842	**29**
30	13.268	.0754	.00734	.0973	136.308	10.274	8.666	89.028	**30**
31	14.462	.0691	.00669	.0967	149.575	10.343	8.808	91.102	**31**
32	15.763	.0634	.00610	.0961	164.037	10.406	8.944	93.069	**32**
33	17.182	.0582	.00556	.0956	179.801	10.464	9.072	94.931	**33**
34	18.728	.0534	.00508	.0951	196.983	10.518	9.193	96.693	**34**
35	20.414	.0490	.00464	.0946	215.711	10.567	9.308	98.359	**35**
40	31.409	.0318	.00296	.0930	337.883	10.757	9.796	105.376	**40**
45	48.327	.0207	.00190	.0919	525.860	10.881	10.160	110.556	**45**
50	74.358	.0134	.00123	.0912	815.085	10.962	10.430	114.325	**50**
55	114.409	.00874	.00079	.0908	1 260.1	11.014	10.626	117.036	**55**
60	176.032	.00568	.00051	.0905	1 944.8	11.048	10.768	118.968	**60**
65	270.847	.00369	.00033	.0903	2 998.3	11.070	10.870	120.334	**65**
70	416.731	.00240	.00022	.0902	4 619.2	11.084	10.943	121.294	**70**
75	641.193	.00156	.00014	.0901	7 113.3	11.094	10.994	121.965	**75**
80	986.555	.00101	.00009	.0901	10 950.6	11.100	11.030	122.431	**80**
85	1 517.9	.00066	.00006	.0901	16 854.9	11.104	11.055	122.753	**85**
90	2 335.5	.00043	.00004	.0900	25 939.3	11.106	11.073	122.976	**90**
95	3 593.5	.00028	.00003	.0900	39 916.8	11.108	11.085	123.129	**95**
100	5 529.1	.00018	.00002	.0900	61 422.9	11.109	11.093	123.233	**100**

n	Single Payment		Uniform Payment Series				Arithmetic Gradient		n
	Compound Amount Factor	Present Worth Factor	Sinking Fund Factor	Capital Recovery Factor	Compound Amount Factor	Present Worth Factor	Gradient Uniform Series	Gradient Present Worth	
	Find *F* Given *P* *F/P*	Find *P* Given *F* *P/F*	Find *A* Given *F* *A/F*	Find *A* Given *P* *A/P*	Find *F* Given *A* *F/A*	Find *P* Given *A* *P/A*	Find *A* Given *G* *A/G*	Find *P* Given *G* *P/G*	
1	1.100	.9091	1.0000	1.1000	1.000	0.909	0	0	**1**
2	1.210	.8264	.4762	.5762	2.100	1.736	0.476	0.826	**2**
3	1.331	.7513	.3021	.4021	3.310	2.487	0.937	2.329	**3**
4	1.464	.6830	.2155	.3155	4.641	3.170	1.381	4.378	**4**
5	1.611	.6209	.1638	.2638	6.105	3.791	1.810	6.862	**5**
6	1.772	.5645	.1296	.2296	7.716	4.355	2.224	9.684	**6**
7	1.949	.5132	.1054	.2054	9.487	4.868	2.622	12.763	**7**
8	2.144	.4665	.0874	.1874	11.436	5.335	3.004	16.029	**8**
9	2.358	.4241	.0736	.1736	13.579	5.759	3.372	19.421	**9**
10	2.594	.3855	.0627	.1627	15.937	6.145	3.725	22.891	**10**
11	2.853	.3505	.0540	.1540	18.531	6.495	4.064	26.396	**11**
12	3.138	.3186	.0468	.1468	21.384	6.814	4.388	29.901	**12**
13	3.452	.2897	.0408	.1408	24.523	7.103	4.699	33.377	**13**
14	3.797	.2633	.0357	.1357	27.975	7.367	4.996	36.801	**14**
15	4.177	.2394	.0315	.1315	31.772	7.606	5.279	40.152	**15**
16	4.595	.2176	.0278	.1278	35.950	7.824	5.549	43.416	**16**
17	5.054	.1978	.0247	.1247	40.545	8.022	5.807	46.582	**17**
18	5.560	.1799	.0219	.1219	45.599	8.201	6.053	49.640	**18**
19	6.116	.1635	.0195	.1195	51.159	8.365	6.286	52.583	**19**
20	6.728	.1486	.0175	.1175	57.275	8.514	6.508	55.407	**20**
21	7.400	.1351	.0156	.1156	64.003	8.649	6.719	58.110	**21**
22	8.140	.1228	.0140	.1140	71.403	8.772	6.919	60.689	**22**
23	8.954	.1117	.0126	.1126	79.543	8.883	7.108	63.146	**23**
24	9.850	.1015	.0113	.1113	88.497	8.985	7.288	65.481	**24**
25	10.835	.0923	.0102	.1102	98.347	9.077	7.458	67.696	**25**
26	11.918	.0839	.00916	.1092	109.182	9.161	7.619	69.794	**26**
27	13.110	.0763	.00826	.1083	121.100	9.237	7.770	71.777	**27**
28	14.421	.0693	.00745	.1075	134.210	9.307	7.914	73.650	**28**
29	15.863	.0630	.00673	.1067	148.631	9.370	8.049	75.415	**29**
30	17.449	.0573	.00608	.1061	164.494	9.427	8.176	77.077	**30**
31	19.194	.0521	.00550	.1055	181.944	9.479	8.296	78.640	**31**
32	21.114	.0474	.00497	.1050	201.138	9.526	8.409	80.108	**32**
33	23.225	.0431	.00450	.1045	222.252	9.569	8.515	81.486	**33**
34	25.548	.0391	.00407	.1041	245.477	9.609	8.615	82.777	**34**
35	28.102	.0356	.00369	.1037	271.025	9.644	8.709	83.987	**35**
40	45.259	.0221	.00226	.1023	442.593	9.779	9.096	88.953	**40**
45	72.891	.0137	.00139	.1014	718.905	9.863	9.374	92.454	**45**
50	117.391	.00852	.00086	.1009	1 163.9	9.915	9.570	94.889	**50**
55	189.059	.00529	.00053	.1005	1 880.6	9.947	9.708	96.562	**55**
60	304.482	.00328	.00033	.1003	3 034.8	9.967	9.802	97.701	**60**
65	490.371	.00204	.00020	.1002	4 893.7	9.980	9.867	98.471	**65**
70	789.748	.00127	.00013	.1001	7 887.5	9.987	9.911	98.987	**70**
75	1 271.9	.00079	.00008	.1001	12 709.0	9.992	9.941	99.332	**75**
80	2 048.4	.00049	.00005	.1000	20 474.0	9.995	9.961	99.561	**80**
85	3 299.0	.00030	.00003	.1000	32 979.7	9.997	9.974	99.712	**85**
90	5 313.0	.00019	.00002	.1000	53 120.3	9.998	9.983	99.812	**90**
95	8 556.7	.00012	.00001	.1000	85 556.9	9.999	9.989	99.877	**95**
100	13 780.6	.00007	.00001	.1000	137 796.3	9.999	9.993	99.920	**100**

12% Compound Interest Factors 12%

	Single Payment		Uniform Payment Series				Arithmetic Gradient		
	Compound Amount Factor	Present Worth Factor	Sinking Fund Factor	Capital Recovery Factor	Compound Amount Factor	Present Worth Factor	Gradient Uniform Series	Gradient Present Worth	
n	Find *F* Given *P* *F/P*	Find *P* Given *F* *P/F*	Find *A* Given *F* *A/F*	Find *A* Given *P* *A/P*	Find *F* Given *A* *F/A*	Find *P* Given *A* *P/A*	Find *A* Given *G* *A/G*	Find *P* Given *G* *P/G*	*n*
1	1.120	.8929	1.0000	1.1200	1.000	0.893	0	0	**1**
2	1.254	.7972	.4717	.5917	2.120	1.690	0.472	0.797	**2**
3	1.405	.7118	.2963	.4163	3.374	2.402	0.925	2.221	**3**
4	1.574	.6355	.2092	.3292	4.779	3.037	1.359	4.127	**4**
5	1.762	.5674	.1574	.2774	6.353	3.605	1.775	6.397	**5**
6	1.974	.5066	.1232	.2432	8.115	4.111	2.172	8.930	**6**
7	2.211	.4523	.0991	.2191	10.089	4.564	2.551	11.644	**7**
8	2.476	.4039	.0813	.2013	12.300	4.968	2.913	14.471	**8**
9	2.773	.3606	.0677	.1877	14.776	5.328	3.257	17.356	**9**
10	3.106	.3220	.0570	.1770	17.549	5.650	3.585	20.254	**10**
11	3.479	.2875	.0484	.1684	20.655	5.938	3.895	23.129	**11**
12	3.896	.2567	.0414	.1614	24.133	6.194	4.190	25.952	**12**
13	4.363	.2292	.0357	.1557	28.029	6.424	4.468	28.702	**13**
14	4.887	.2046	.0309	.1509	32.393	6.628	4.732	31.362	**14**
15	5.474	.1827	.0268	.1468	37.280	6.811	4.980	33.920	**15**
16	6.130	.1631	.0234	.1434	42.753	6.974	5.215	36.367	**16**
17	6.866	.1456	.0205	.1405	48.884	7.120	5.435	38.697	**17**
18	7.690	.1300	.0179	.1379	55.750	7.250	5.643	40.908	**18**
19	8.613	.1161	.0158	.1358	63.440	7.366	5.838	42.998	**19**
20	9.646	.1037	.0139	.1339	72.052	7.469	6.020	44.968	**20**
21	10.804	.0926	.0122	.1322	81.699	7.562	6.191	46.819	**21**
22	12.100	.0826	.0108	.1308	92.503	7.645	6.351	48.554	**22**
23	13.552	.0738	.00956	.1296	104.603	7.718	6.501	50.178	**23**
24	15.179	.0659	.00846	.1285	118.155	7.784	6.641	51.693	**24**
25	17.000	.0588	.00750	.1275	133.334	7.843	6.771	53.105	**25**
26	19.040	.0525	.00665	.1267	150.334	7.896	6.892	54.418	**26**
27	21.325	.0469	.00590	.1259	169.374	7.943	7.005	55.637	**27**
28	23.884	.0419	.00524	.1252	190.699	7.984	7.110	56.767	**28**
29	26.750	.0374	.00466	.1247	214.583	8.022	7.207	57.814	**29**
30	29.960	.0334	.00414	.1241	241.333	8.055	7.297	58.782	**30**
31	33.555	.0298	.00369	.1237	271.293	8.085	7.381	59.676	**31**
32	37.582	.0266	.00328	.1233	304.848	8.112	7.459	60.501	**32**
33	42.092	.0238	.00292	.1229	342.429	8.135	7.530	61.261	**33**
34	47.143	.0212	.00260	.1226	384.521	8.157	7.596	61.961	**34**
35	52.800	.0189	.00232	.1223	431.663	8.176	7.658	62.605	**35**
40	93.051	.0107	.00130	.1213	767.091	8.244	7.899	65.116	**40**
45	163.988	.00610	.00074	.1207	1 358.2	8.283	8.057	66.734	**45**
50	289.002	.00346	.00042	.1204	2 400.0	8.304	8.160	67.762	**50**
55	509.321	.00196	.00024	.1202	4 236.0	8.317	8.225	68.408	**55**
60	897.597	.00111	.00013	.1201	7 471.6	8.324	8.266	68.810	**60**
65	1 581.9	.00063	.00008	.1201	13 173.9	8.328	8.292	69.058	**65**
70	2 787.8	.00036	.00004	.1200	23 223.3	8.330	8.308	69.210	**70**
75	4 913.1	.00020	.00002	.1200	40 933.8	8.332	8.318	69.303	**75**
80	8 658.5	.00012	.00001	.1200	72 145.7	8.332	8.324	69.359	**80**
85	15 259.2	.00007	.00001	.1200	127 151.7	8.333	8.328	69.393	**85**
90	26 891.9	.00004		.1200	224 091.1	8.333	8.330	69.414	**90**
95	47 392.8	.00002		.1200	394 931.4	8.333	8.331	69.426	**95**
100	83 522.3	.00001		.1200	696 010.5	8.333	8.332	69.434	**100**

	Single Payment		Uniform Payment Series				Arithmetic Gradient		
	Compound Amount Factor	Present Worth Factor	Sinking Fund Factor	Capital Recovery Factor	Compound Amount Factor	Present Worth Factor	Gradient Uniform Series	Gradient Present Worth	
n	Find *F* Given *P* *F/P*	Find *P* Given *F* *P/F*	Find *A* Given *F* *A/F*	Find *A* Given *P* *A/P*	Find *F* Given *A* *F/A*	Find *P* Given *A* *P/A*	Find *A* Given *G* *A/G*	Find *P* Given *G* *P/G*	n
1	1.150	.8696	1.0000	1.1500	1.000	0.870	0	0	**1**
2	1.322	.7561	.4651	.6151	2.150	1.626	0.465	0.756	**2**
3	1.521	.6575	.2880	.4380	3.472	2.283	0.907	2.071	**3**
4	1.749	.5718	.2003	.3503	4.993	2.855	1.326	3.786	**4**
5	2.011	.4972	.1483	.2983	6.742	3.352	1.723	5.775	**5**
6	2.313	.4323	.1142	.2642	8.754	3.784	2.097	7.937	**6**
7	2.660	.3759	.0904	.2404	11.067	4.160	2.450	10.192	**7**
8	3.059	.3269	.0729	.2229	13.727	4.487	2.781	12.481	**8**
9	3.518	.2843	.0596	.2096	16.786	4.772	3.092	14.755	**9**
10	4.046	.2472	.0493	.1993	20.304	5.019	3.383	16.979	**10**
11	4.652	.2149	.0411	.1911	24.349	5.234	3.655	19.129	**11**
12	5.350	.1869	.0345	.1845	29.002	5.421	3.908	21.185	**12**
13	6.153	.1625	.0291	.1791	34.352	5.583	4.144	23.135	**13**
14	7.076	.1413	.0247	.1747	40.505	5.724	4.362	24.972	**14**
15	8.137	.1229	.0210	.1710	47.580	5.847	4.565	26.693	**15**
16	9.358	.1069	.0179	.1679	55.717	5.954	4.752	28.296	**16**
17	10.761	.0929	.0154	.1654	65.075	6.047	4.925	29.783	**17**
18	12.375	.0808	.0132	.1632	75.836	6.128	5.084	31.156	**18**
19	14.232	.0703	.0113	.1613	88.212	6.198	5.231	32.421	**19**
20	16.367	.0611	.00976	.1598	102.444	6.259	5.365	33.582	**20**
21	18.822	.0531	.00842	.1584	118.810	6.312	5.488	34.645	**21**
22	21.645	.0462	.00727	.1573	137.632	6.359	5.601	35.615	**22**
23	24.891	.0402	.00628	.1563	159.276	6.399	5.704	36.499	**23**
24	28.625	.0349	.00543	.1554	184.168	6.434	5.798	37.302	**24**
25	32.919	.0304	.00470	.1547	212.793	6.464	5.883	38.031	**25**
26	37.857	.0264	.00407	.1541	245.712	6.491	5.961	38.692	**26**
27	43.535	.0230	.00353	.1535	283.569	6.514	6.032	39.289	**27**
28	50.066	.0200	.00306	.1531	327.104	6.534	6.096	39.828	**28**
29	57.575	.0174	.00265	.1527	377.170	6.551	6.154	40.315	**29**
30	66.212	.0151	.00230	.1523	434.745	6.566	6.207	40.753	**30**
31	76.144	.0131	.00200	.1520	500.957	6.579	6.254	41.147	**31**
32	87.565	.0114	.00173	.1517	577.100	6.591	6.297	41.501	**32**
33	100.700	.00993	.00150	.1515	664.666	6.600	6.336	41.818	**33**
34	115.805	.00864	.00131	.1513	765.365	6.609	6.371	42.103	**34**
35	133.176	.00751	.00113	.1511	881.170	6.617	6.402	42.359	**35**
40	267.864	.00373	.00056	.1506	1 779.1	6.642	6.517	43.283	**40**
45	538.769	.00186	.00028	.1503	3 585.1	6.654	6.583	43.805	**45**
50	1 083.7	.00092	.00014	.1501	7 217.7	6.661	6.620	44.096	**50**
55	2 179.6	.00046	.00007	.1501	14 524.1	6.664	6.641	44.256	**55**
60	4 384.0	.00023	.00003	.1500	29 220.0	6.665	6.653	44.343	**60**
65	8 817.8	.00011	.00002	.1500	58 778.6	6.666	6.659	44.390	**65**
70	17 735.7	.00006	.00001	.1500	118 231.5	6.666	6.663	44.416	**70**
75	35 672.9	.00003		.1500	237 812.5	6.666	6.665	44.429	**75**
80	71 750.9	.00001		.1500	478 332.6	6.667	6.666	44.436	**80**
85	144 316.7	.00001		.1500	962 104.4	6.667	6.666	44.440	**85**

Compound Interest Factors

18%

	Single Payment		Uniform Payment Series				Arithmetic Gradient		
	Compound Amount Factor	Present Worth Factor	Sinking Fund Factor	Capital Recovery Factor	Compound Amount Factor	Present Worth Factor	Gradient Uniform Series	Gradient Present Worth	
n	Find *F* Given *P* *F/P*	Find *P* Given *F* *P/F*	Find *A* Given *F* *A/F*	Find *A* Given *P* *A/P*	Find *F* Given *A* *F/A*	Find *P* Given *A* *P/A*	Find *A* Given *G* *A/G*	Find *P* Given *G* *P/G*	*n*
1	1.180	.8475	1.0000	1.1800	1.000	0.847	0	0	**1**
2	1.392	.7182	.4587	.6387	2.180	1.566	0.459	0.718	**2**
3	1.643	.6086	.2799	.4599	3.572	2.174	0.890	1.935	**3**
4	1.939	.5158	.1917	.3717	5.215	2.690	1.295	3.483	**4**
5	2.288	.4371	.1398	.3198	7.154	3.127	1.673	5.231	**5**
6	2.700	.3704	.1059	.2859	9.442	3.498	2.025	7.083	**6**
7	3.185	.3139	.0824	.2624	12.142	3.812	2.353	8.967	**7**
8	3.759	.2660	.0652	.2452	15.327	4.078	2.656	10.829	**8**
9	4.435	.2255	.0524	.2324	19.086	4.303	2.936	12.633	**9**
10	5.234	.1911	.0425	.2225	23.521	4.494	3.194	14.352	**10**
11	6.176	.1619	.0348	.2148	28.755	4.656	3.430	15.972	**11**
12	7.288	.1372	.0286	.2086	34.931	4.793	3.647	17.481	**12**
13	8.599	.1163	.0237	.2037	42.219	4.910	3.845	18.877	**13**
14	10.147	.0985	.0197	.1997	50.818	5.008	4.025	20.158	**14**
15	11.974	.0835	.0164	.1964	60.965	5.092	4.189	21.327	**15**
16	14.129	.0708	.0137	.1937	72.939	5.162	4.337	22.389	**16**
17	16.672	.0600	.0115	.1915	87.068	5.222	4.471	23.348	**17**
18	19.673	.0508	.00964	.1896	103.740	5.273	4.592	24.212	**18**
19	23.214	.0431	.00810	.1881	123.413	5.316	4.700	24.988	**19**
20	27.393	.0365	.00682	.1868	146.628	5.353	4.798	25.681	**20**
21	32.324	.0309	.00575	.1857	174.021	5.384	4.885	26.300	**21**
22	38.142	.0262	.00485	.1848	206.345	5.410	4.963	26.851	**22**
23	45.008	.0222	.00409	.1841	244.487	5.432	5.033	27.339	**23**
24	53.109	.0188	.00345	.1835	289.494	5.451	5.095	27.772	**24**
25	62.669	.0160	.00292	.1829	342.603	5.467	5.150	28.155	**25**
26	73.949	.0135	.00247	.1825	405.272	5.480	5.199	28.494	**26**
27	87.260	.0115	.00209	.1821	479.221	5.492	5.243	28.791	**27**
28	102.966	.00971	.00177	.1818	566.480	5.502	5.281	29.054	**28**
29	121.500	.00823	.00149	.1815	669.447	5.510	5.315	29.284	**29**
30	143.370	.00697	.00126	.1813	790.947	5.517	5.345	29.486	**30**
31	169.177	.00591	.00107	.1811	934.317	5.523	5.371	29.664	**31**
32	199.629	.00501	.00091	.1809	1 103.5	5.528	5.394	29.819	**32**
33	235.562	.00425	.00077	.1808	1 303.1	5.532	5.415	29.955	**33**
34	277.963	.00360	.00065	.1806	1 538.7	5.536	5.433	30.074	**34**
35	327.997	.00305	.00055	.1806	1 816.6	5.539	5.449	30.177	**35**
40	750.377	.00133	.00024	.1802	4 163.2	5.548	5.502	30.527	**40**
45	1 716.7	.00058	.00010	.1801	9 531.6	5.552	5.529	30.701	**45**
50	3 927.3	.00025	.00005	.1800	21 813.0	5.554	5.543	30.786	**50**
55	8 984.8	.00011	.00002	.1800	49 910.1	5.555	5.549	30.827	**55**
60	20 555.1	.00005	.00001	.1800	114 189.4	5.555	5.553	30.846	**60**
65	47 025.1	.00002		.1800	261 244.7	5.555	5.554	30.856	**65**
70	107 581.9	.00001		.1800	597.671.7	5.556	5.555	30.860	**70**

	Single Payment		Uniform Payment Series				Arithmetic Gradient		
	Compound Amount Factor	Present Worth Factor	Sinking Fund Factor	Capital Recovery Factor	Compound Amount Factor	Present Worth Factor	Gradient Uniform Series	Gradient Present Worth	
n	Find *F* Given *P* *F/P*	Find *P* Given *F* *P/F*	Find *A* Given *F* *A/F*	Find *A* Given *P* *A/P*	Find *F* Given *A* *F/A*	Find *P* Given *A* *P/A*	Find *A* Given *G* *A/G*	Find *P* Given *G* *P/G*	*n*
1	1.200	.8333	1.0000	1.2000	1.000	0.833	0	0	**1**
2	1.440	.6944	.4545	.6545	2.200	1.528	0.455	0.694	**2**
3	1.728	.5787	.2747	.4747	3.640	2.106	0.879	1.852	**3**
4	2.074	.4823	.1863	.3863	5.368	2.589	1.274	3.299	**4**
5	2.488	.4019	.1344	.3344	7.442	2.991	1.641	4.906	**5**
6	2.986	.3349	.1007	.3007	9.930	3.326	1.979	6.581	**6**
7	3.583	.2791	.0774	.2774	12.916	3.605	2.290	8.255	**7**
8	4.300	.2326	.0606	.2606	16.499	3.837	2.576	9.883	**8**
9	5.160	.1938	.0481	.2481	20.799	4.031	2.836	11.434	**9**
10	6.192	.1615	.0385	.2385	25.959	4.192	3.074	12.887	**10**
11	7.430	.1346	.0311	.2311	32.150	4.327	3.289	14.233	**11**
12	8.916	.1122	.0253	.2253	39.581	4.439	3.484	15.467	**12**
13	10.699	.0935	.0206	.2206	48.497	4.533	3.660	16.588	**13**
14	12.839	.0779	.0169	.2169	59.196	4.611	3.817	17.601	**14**
15	15.407	.0649	.0139	.2139	72.035	4.675	3.959	18.509	**15**
16	18.488	.0541	.0114	.2114	87.442	4.730	4.085	19.321	**16**
17	22.186	.0451	.00944	.2094	105.931	4.775	4.198	20.042	**17**
18	26.623	.0376	.00781	.2078	128.117	4.812	4.298	20.680	**18**
19	31.948	.0313	.00646	.2065	154.740	4.843	4.386	21.244	**19**
20	38.338	.0261	.00536	.2054	186.688	4.870	4.464	21.739	**20**
21	46.005	.0217	.00444	.2044	225.026	4.891	4.533	22.174	**21**
22	55.206	.0181	.00369	.2037	271.031	4.909	4.594	22.555	**22**
23	66.247	.0151	.00307	.2031	326.237	4.925	4.647	22.887	**23**
24	79.497	.0126	.00255	.2025	392.484	4.937	4.694	23.176	**24**
25	95.396	.0105	.00212	.2021	471.981	4.948	4.735	23.428	**25**
26	114.475	.00874	.00176	.2018	567.377	4.956	4.771	23.646	**26**
27	137.371	.00728	.00147	.2015	681.853	4.964	4.802	23.835	**27**
28	164.845	.00607	.00122	.2012	819.223	4.970	4.829	23.999	**28**
29	197.814	.00506	.00102	.2010	984.068	4.975	4.853	24.141	**29**
30	237.376	.00421	.00085	.2008	1 181.9	4.979	4.873	24.263	**30**
31	284.852	.00351	.00070	.2007	1 419.3	4.982	4.891	24.368	**31**
32	341.822	.00293	.00059	.2006	1 704.1	4.985	4.906	24.459	**32**
33	410.186	.00244	.00049	.2005	2 045.9	4.988	4.919	24.537	**33**
34	492.224	.00203	.00041	.2004	2 456.1	4.990	4.931	24.604	**34**
35	590.668	.00169	.00034	.2003	2 948.3	4.992	4.941	24.661	**35**
40	1 469.8	.00068	.00014	.2001	7 343.9	4.997	4.973	24.847	**40**
45	3 657.3	.00027	.00005	.2001	18 281.3	4.999	4.988	24.932	**45**
50	9 100.4	.00011	.00002	.2000	45 497.2	4.999	4.995	24.970	**50**
55	22 644.8	.00004	.00001	.2000	113 219.0	5.000	4.998	24.987	**55**
60	56 347.5	.00002		.2000	281 732.6	5.000	4.999	24.994	**60**

25% Compound Interest Factors 25%

n	Single Payment		Uniform Payment Series				Arithmetic Gradient		n
	Compound Amount Factor	Present Worth Factor	Sinking Fund Factor	Capital Recovery Factor	Compound Amount Factor	Present Worth Factor	Gradient Uniform Series	Gradient Present Worth	
	Find *F* Given *P* *F/P*	Find *P* Given *F* *P/F*	Find *A* Given *F* *A/F*	Find *A* Given *P* *A/P*	Find *F* Given *A* *F/A*	Find *P* Given *A* *P/A*	Find *A* Given *G* *A/G*	Find *P* Given *G* *P/G*	
1	1.250	.8000	1.0000	1.2500	1.000	0.800	0	0	**1**
2	1.563	.6400	.4444	.6944	2.250	1.440	0.444	0.640	**2**
3	1.953	.5120	.2623	.5123	3.813	1.952	0.852	1.664	**3**
4	2.441	.4096	.1734	.4234	5.766	2.362	1.225	2.893	**4**
5	3.052	.3277	.1218	.3718	8.207	2.689	1.563	4.204	**5**
6	3.815	.2621	.0888	.3388	11.259	2.951	1.868	5.514	**6**
7	4.768	.2097	.0663	.3163	15.073	3.161	2.142	6.773	**7**
8	5.960	.1678	.0504	.3004	19.842	3.329	2.387	7.947	**8**
9	7.451	.1342	.0388	.2888	25.802	3.463	2.605	9.021	**9**
10	9.313	.1074	.0301	.2801	33.253	3.571	2.797	9.987	**10**
11	11.642	.0859	.0235	.2735	42.566	3.656	2.966	10.846	**11**
12	14.552	.0687	.0184	.2684	54.208	3.725	3.115	11.602	**12**
13	18.190	.0550	.0145	.2645	68.760	3.780	3.244	12.262	**13**
14	22.737	.0440	.0115	.2615	86.949	3.824	3.356	12.833	**14**
15	28.422	.0352	.00912	.2591	109.687	3.859	3.453	13.326	**15**
16	35.527	.0281	.00724	.2572	138.109	3.887	3.537	13.748	**16**
17	44.409	.0225	.00576	.2558	173.636	3.910	3.608	14.108	**17**
18	55.511	.0180	.00459	.2546	218.045	3.928	3.670	14.415	**18**
19	69.389	.0144	.00366	.2537	273.556	3.942	3.722	14.674	**19**
20	86.736	.0115	.00292	.2529	342.945	3.954	3.767	14.893	**20**
21	108.420	.00922	.00233	.2523	429.681	3.963	3.805	15.078	**21**
22	135.525	.00738	.00186	.2519	538.101	3.970	3.836	15.233	**22**
23	169.407	.00590	.00148	.2515	673.626	3.976	3.863	15.362	**23**
24	211.758	.00472	.00119	.2512	843.033	3.981	3.886	15.471	**24**
25	264.698	.00378	.00095	.2509	1 054.8	3.985	3.905	15.562	**25**
26	330.872	.00302	.00076	.2508	1 319.5	3.988	3.921	15.637	**26**
27	413.590	.00242	.00061	.2506	1 650.4	3.990	3.935	15.700	**27**
28	516.988	.00193	.00048	.2505	2 064.0	3.992	3.946	15.752	**28**
29	646.235	.00155	.00039	.2504	2 580.9	3.994	3.955	15.796	**29**
30	807.794	.00124	.00031	.2503	3 227.2	3.995	3.963	15.832	**30**
31	1 009.7	.00099	.00025	.2502	4 035.0	3.996	3.969	15.861	**31**
32	1 262.2	.00079	.00020	.2502	5 044.7	3.997	3.975	15.886	**32**
33	1 577.7	.00063	.00016	.2502	6 306.9	3.997	3.979	15.906	**33**
34	1 972.2	.00051	.00013	.2501	7 884.6	3.998	3.983	15.923	**34**
35	2 465.2	.00041	.00010	.2501	9 856.8	3.998	3.986	15.937	**35**
40	7 523.2	.00013	.00003	.2500	30 088.7	3.999	3.995	15.977	**40**
45	22 958.9	.00004	.00001	.2500	91 831.5	4.000	3.998	15.991	**45**
50	70 064.9	.00001		.2500	280 255.7	4.000	3.999	15.997	**50**
55	213 821.2			.2500	855 280.7	4.000	4.000	15.999	**55**

	Single Payment		Uniform Payment Series				Arithmetic Gradient		
	Compound Amount Factor	Present Worth Factor	Sinking Fund Factor	Capital Recovery Factor	Compound Amount Factor	Present Worth Factor	Gradient Uniform Series	Gradient Present Worth	
n	Find *F* Given *P* *F/P*	Find *P* Given *F* *P/F*	Find *A* Given *F* *A/F*	Find *A* Given *P* *A/P*	Find *F* Given *A* *F/A*	Find *P* Given *A* *P/A*	Find *A* Given *G* *A/G*	Find *P* Given *G* *P/G*	**n**
1	1.300	.7692	1.0000	1.3000	1.000	0.769	0	0	**1**
2	1.690	.5917	.4348	.7348	2.300	1.361	0.435	0.592	**2**
3	2.197	.4552	.2506	.5506	3.990	1.816	0.827	1.502	**3**
4	2.856	.3501	.1616	.4616	6.187	2.166	1.178	2.552	**4**
5	3.713	.2693	.1106	.4106	9.043	2.436	1.490	3.630	**5**
6	4.827	.2072	.0784	.3784	12.756	2.643	1.765	4.666	**6**
7	6.275	.1594	.0569	.3569	17.583	2.802	2.006	5.622	**7**
8	8.157	.1226	.0419	.3419	23.858	2.925	2.216	6.480	**8**
9	10.604	.0943	.0312	.3312	32.015	3.019	2.396	7.234	**9**
10	13.786	.0725	.0235	.3235	42.619	3.092	2.551	7.887	**10**
11	17.922	.0558	.0177	.3177	56.405	3.147	2.683	8.445	**11**
12	23.298	.0429	.0135	.3135	74.327	3.190	2.795	8.917	**12**
13	30.287	.0330	.0102	.3102	97.625	3.223	2.889	9.314	**13**
14	39.374	.0254	.00782	.3078	127.912	3.249	2.969	9.644	**14**
15	51.186	.0195	.00598	.3060	167.286	3.268	3.034	9.917	**15**
16	66.542	.0150	.00458	.3046	218.472	3.283	3.089	10.143	**16**
17	86.504	.0116	.00351	.3035	285.014	3.295	3.135	10.328	**17**
18	112.455	.00889	.00269	.3027	371.518	3.304	3.172	10.479	**18**
19	146.192	.00684	.00207	.3021	483.973	3.311	3.202	10.602	**19**
20	190.049	.00526	.00159	.3016	630.165	3.316	3.228	10.702	**20**
21	247.064	.00405	.00122	.3012	820.214	3.320	3.248	10.783	**21**
22	321.184	.00311	.00094	.3009	1 067.3	3.323	3.265	10.848	**22**
23	417.539	.00239	.00072	.3007	1 388.5	3.325	3.278	10.901	**23**
24	542.800	.00184	.00055	.3006	1 806.0	3.327	3.289	10.943	**24**
25	705.640	.00142	.00043	.3004	2 348.8	3.329	3.298	10.977	**25**
26	917.332	.00109	.00033	.3003	3 054.4	3.330	3.305	11.005	**26**
27	1 192.5	.00084	.00025	.3003	3 971.8	3.331	3.311	11.026	**27**
28	1 550.3	.00065	.00019	.3002	5 164.3	3.331	3.315	11.044	**28**
29	2 015.4	.00050	.00015	.3001	6 714.6	3.332	3.319	11.058	**29**
30	2 620.0	.00038	.00011	.3001	8 730.0	3.332	3.322	11.069	**30**
31	3 406.0	.00029	.00009	.3001	11 350.0	3.332	3.324	11.078	**31**
32	4 427.8	.00023	.00007	.3001	14 756.0	3.333	3.326	11.085	**32**
33	5 756.1	.00017	.00005	.3001	19 183.7	3.333	3.328	11.090	**33**
34	7 483.0	.00013	.00004	.3000	24 939.9	3.333	3.329	11.094	**34**
35	9 727.8	.00010	.00003	.3000	32 422.8	3.333	3.330	11.098	**35**
40	36 118.8	.00003	.00001	.3000	120 392.6	3.333	3.332	11.107	**40**
45	134 106.5	.00001		.3000	447 018.3	3.333	3.333	11.110	**45**

n	Single Payment		Uniform Payment Series				Arithmetic Gradient		n
	Compound Amount Factor	Present Worth Factor	Sinking Fund Factor	Capital Recovery Factor	Compound Amount Factor	Present Worth Factor	Gradient Uniform Series	Gradient Present Worth	
	Find F Given P F/P	Find P Given F P/F	Find A Given F A/F	Find A Given P A/P	Find F Given A F/A	Find P Given A P/A	Find A Given G A/G	Find P Given G P/G	
1	1.350	.7407	1.0000	1.3500	1.000	0.741	0	0	**1**
2	1.822	.5487	.4255	.7755	2.350	1.289	0.426	0.549	**2**
3	2.460	.4064	.2397	.5897	4.173	1.696	0.803	1.362	**3**
4	3.322	.3011	.1508	.5008	6.633	1.997	1.134	2.265	**4**
5	4.484	.2230	.1005	.4505	9.954	2.220	1.422	3.157	**5**
6	6.053	.1652	.0693	.4193	14.438	2.385	1.670	3.983	**6**
7	8.172	.1224	.0488	.3988	20.492	2.508	1.881	4.717	**7**
8	11.032	.0906	.0349	.3849	28.664	2.598	2.060	5.352	**8**
9	14.894	.0671	.0252	.3752	39.696	2.665	2.209	5.889	**9**
10	20.107	.0497	.0183	.3683	54.590	2.715	2.334	6.336	**10**
11	27.144	.0368	.0134	.3634	74.697	2.752	2.436	6.705	**11**
12	36.644	.0273	.00982	.3598	101.841	2.779	2.520	7.005	**12**
13	49.470	.0202	.00722	.3572	138.485	2.799	2.589	7.247	**13**
14	66.784	.0150	.00532	.3553	187.954	2.814	2.644	7.442	**14**
15	90.158	.0111	.00393	.3539	254.739	2.825	2.689	7.597	**15**
16	121.714	.00822	.00290	.3529	344.897	2.834	2.725	7.721	**16**
17	164.314	.00609	.00214	.3521	466.611	2.840	2.753	7.818	**17**
18	221.824	.00451	.00158	.3516	630.925	2.844	2.776	7.895	**18**
19	299.462	.00334	.00117	.3512	852.748	2.848	2.793	7.955	**19**
20	404.274	.00247	.00087	.3509	1 152.2	2.850	2.808	8.002	**20**
21	545.769	.00183	.00064	.3506	1 556.5	2.852	2.819	8.038	**21**
22	736.789	.00136	.00048	.3505	2 102.3	2.853	2.827	8.067	**22**
23	994.665	.00101	.00035	.3504	2 839.0	2.854	2.834	8.089	**23**
24	1 342.8	.00074	.00026	.3503	3 833.7	2.855	2.839	8.106	**24**
25	1 812.8	.00055	.00019	.3502	5 176.5	2.856	2.843	8.119	**25**
26	2 447.2	.00041	.00014	.3501	6 989.3	2.856	2.847	8.130	**26**
27	3 303.8	.00030	.00011	.3501	9 436.5	2.856	2.849	8.137	**27**
28	4 460.1	.00022	.00008	.3501	12 740.3	2.857	2.851	8.143	**28**
29	6 021.1	.00017	.00006	.3501	17 200.4	2.857	2.852	8.148	**29**
30	8 128.5	.00012	.00004	.3500	23 221.6	2.857	2.853	8.152	**30**
31	10 973.5	.00009	.00003	.3500	31 350.1	2.857	2.854	8.154	**31**
32	14 814.3	.00007	.00002	.3500	42 323.7	2.857	2.855	8.157	**32**
33	19 999.3	.00005	.00002	.3500	57 137.9	2.857	2.855	8.158	**33**
34	26 999.0	.00004	.00001	.3500	77 137.2	2.857	2.856	8.159	**34**
35	36 448.7	.00003	.00001	.3500	104 136.3	2.857	2.856	8.160	**35**

40% Compound Interest Factors 40%

	Single Payment		Uniform Payment Series				Arithmetic Gradient		
	Compound Amount Factor	Present Worth Factor	Sinking Fund Factor	Capital Recovery Factor	Compound Amount Factor	Present Worth Factor	Gradient Uniform Series	Gradient Present Worth	
n	Find *F* Given *P* *F/P*	Find *P* Given *F* *P/F*	Find *A* Given *F* *A/F*	Find *A* Given *P* *A/P*	Find *F* Given *A* *F/A*	Find *P* Given *A* *P/A*	Find *A* Given *G* *A/G*	Find *P* Given *G* *P/G*	**n**
1	1.400	.7143	1.0000	1.4000	1.000	0.714	0	0	**1**
2	1.960	.5102	.4167	.8167	2.400	1.224	0.417	0.510	**2**
3	2.744	.3644	.2294	.6294	4.360	1.589	0.780	1.239	**3**
4	3.842	.2603	.1408	.5408	7.104	1.849	1.092	2.020	**4**
5	5.378	.1859	.0914	.4914	10.946	2.035	1.358	2.764	**5**
6	7.530	.1328	.0613	.4613	16.324	2.168	1.581	3.428	**6**
7	10.541	.0949	.0419	.4419	23.853	2.263	1.766	3.997	**7**
8	14.758	.0678	.0291	.4291	34.395	2.331	1.919	4.471	**8**
9	20.661	.0484	.0203	.4203	49.153	2.379	2.042	4.858	**9**
10	28.925	.0346	.0143	.4143	69.814	2.414	2.142	5.170	**10**
11	40.496	.0247	.0101	.4101	98.739	2.438	2.221	5.417	**11**
12	56.694	.0176	.00718	.4072	139.235	2.456	2.285	5.611	**12**
13	79.371	.0126	.00510	.4051	195.929	2.469	2.334	5.762	**13**
14	111.120	.00900	.00363	.4036	275.300	2.478	2.373	5.879	**14**
15	155.568	.00643	.00259	.4026	386.420	2.484	2.403	5.969	**15**
16	217.795	.00459	.00185	.4018	541.988	2.489	2.426	6.038	**16**
17	304.913	.00328	.00132	.4013	759.783	2.492	2.444	6.090	**17**
18	426.879	.00234	.00094	.4009	1 064.7	2.494	2.458	6.130	**18**
19	597.630	.00167	.00067	.4007	1 419.6	2.496	2.468	6.160	**19**
20	836.682	.00120	.00048	.4005	2 089.2	2.497	2.476	6.183	**20**
21	1 171.4	.00085	.00034	.4003	2 925.9	2.498	2.482	6.200	**21**
22	1 639.9	.00061	.00024	.4002	4 097.2	2.498	2.487	6.213	**22**
23	2 295.9	.00044	.00017	.4002	5 737.1	2.499	2.490	6.222	**23**
24	3 214.2	.00031	.00012	.4001	8 033.0	2.499	2.493	6.229	**24**
25	4 499.9	.00022	.00009	.4001	11 247.2	2.499	2.494	6.235	**25**
26	6 299.8	.00016	.00006	.4001	15 747.1	2.500	2.496	6.239	**26**
27	8 819.8	.00011	.00005	.4000	22 046.9	2.500	2.497	6.242	**27**
28	12 347.7	.00008	.00003	.4000	30 866.7	2.500	2.498	6.244	**28**
29	17 286.7	.00006	.00002	.4000	43 214.3	2.500	2.498	6.245	**29**
30	24 201.4	.00004	.00002	.4000	60 501.0	2.500	2.499	6.247	**30**
31	33 882.0	.00003	.00001	.4000	84 702.5	2.500	2.499	6.248	**31**
32	47 434.8	.00002	.00001	.4000	118 584.4	2.500	2.499	6.248	**32**
33	66 408.7	.00002	.00001	.4000	166 019.2	2.500	2.500	6.249	**33**
34	92 972.1	.00001		.4000	232 427.9	2.500	2.500	6.249	**34**
35	130 161.0	.00001		.4000	325 400.0	2.500	2.500	6.249	**35**

Compound Interest Factors

45%

	Single Payment		Uniform Payment Series				Arithmetic Gradient		
	Compound Amount Factor	Present Worth Factor	Sinking Fund Factor	Capital Recovery Factor	Compound Amount Factor	Present Worth Factor	Gradient Uniform Series	Gradient Present Worth	
n	Find *F* Given *P* *F/P*	Find *P* Given *F* *P/F*	Find *A* Given *F* *A/F*	Find *A* Given *P* *A/P*	Find *F* Given *A* *F/A*	Find *P* Given *A* *P/A*	Find *A* Given *G* *A/G*	Find *P* Given *G* *P/G*	*n*
1	1.450	.6897	1.0000	1.4500	1.000	0.690	0	0	**1**
2	2.103	.4756	.4082	.8582	2.450	1.165	0.408	0.476	**2**
3	3.049	.3280	.2197	.6697	4.553	1.493	0.758	1.132	**3**
4	4.421	.2262	.1316	.5816	7.601	1.720	1.053	1.810	**4**
5	6.410	.1560	.0832	.5332	12.022	1.876	1.298	2.434	**5**
6	9.294	.1076	.0543	.5043	18.431	1.983	1.499	2.972	**6**
7	13.476	.0742	.0361	.4861	27.725	2.057	1.661	3.418	**7**
8	19.541	.0512	.0243	.4743	41.202	2.109	1.791	3.776	**8**
9	28.334	.0353	.0165	.4665	60.743	2.144	1.893	4.058	**9**
10	41.085	.0243	.0112	.4612	89.077	2.168	1.973	4.277	**10**
11	59.573	.0168	.00768	.4577	130.162	2.185	2.034	4.445	**11**
12	86.381	.0116	.00527	.4553	189.735	2.196	2.082	4.572	**12**
13	125.252	.00798	.00362	.4536	276.115	2.204	2.118	4.668	**13**
14	181.615	.00551	.00249	.4525	401.367	2.210	2.145	4.740	**14**
15	263.342	.00380	.00172	.4517	582.982	2.214	2.165	4.793	**15**
16	381.846	.00262	.00118	.4512	846.325	2.216	2.180	4.832	**16**
17	553.677	.00181	.00081	.4508	1 228.2	2.218	2.191	4.861	**17**
18	802.831	.00125	.00056	.4506	1 781.8	2.219	2.200	4.882	**18**
19	1 164.1	.00086	.00039	.4504	2 584.7	2.220	2.206	4.898	**19**
20	1 688.0	.00059	.00027	.4503	3 748.8	2.221	2.210	4.909	**20**
21	2 447.5	.00041	.00018	.4502	5 436.7	2.221	2.214	4.917	**21**
22	3 548.9	.00028	.00013	.4501	7 884.3	2.222	2.216	4.923	**22**
23	5 145.9	.00019	.00009	.4501	11 433.2	2.222	2.218	4.927	**23**
24	7 461.6	.00013	.00006	.4501	16 579.1	2.222	2.219	4.930	**24**
25	10 819.3	.00009	.00004	.4500	24 040.7	2.222	2.220	4.933	**25**
26	15 688.0	.00006	.00003	.4500	34 860.1	2.222	2.221	4.934	**26**
27	22 747.7	.00004	.00002	.4500	50 548.1	2.222	2.221	4.935	**27**
28	32 984.1	.00003	.00001	.4500	73 295.8	2.222	2.221	4.936	**28**
29	47 826.9	.00002	.00001	.4500	106 279.9	2.222	2.222	4.937	**29**
30	69 349.1	.00001	.00001	.4500	154 106.8	2.222	2.222	4.937	**30**
31	100 556.1	.00001		.4500	223 455.9	2.222	2.222	4.938	**31**
32	145 806.4	.00001		.4500	324 012.0	2.222	2.222	4.938	**32**
33	211 419.3			.4500	469 818.5	2.222	2.222	4.938	**33**
34	306 558.0			.4500	681 237.8	2.222	2.222	4.938	**34**
35	444 509.2			.4500	987 795.9	2.222	2.222	4.938	**35**

n	Single Payment: Compound Amount Factor, Find F Given P, F/P	Single Payment: Present Worth Factor, Find P Given F, P/F	Uniform Payment Series: Sinking Fund Factor, Find A Given F, A/F	Uniform Payment Series: Capital Recovery Factor, Find A Given P, A/P	Uniform Payment Series: Compound Amount Factor, Find F Given A, F/A	Uniform Payment Series: Present Worth Factor, Find P Given A, P/A	Arithmetic Gradient: Gradient Uniform Series, Find A Given G, A/G	Arithmetic Gradient: Gradient Present Worth, Find P Given G, P/G	n
1	1.500	.6667	1.0000	1.5000	1.000	0.667	0	0	1
2	2.250	.4444	.4000	.9000	2.500	1.111	0.400	0.444	2
3	3.375	.2963	.2105	.7105	4.750	1.407	0.737	1.037	3
4	5.063	.1975	.1231	.6231	8.125	1.605	1.015	1.630	4
5	7.594	.1317	.0758	.5758	13.188	1.737	1.242	2.156	5
6	11.391	.0878	.0481	.5481	20.781	1.824	1.423	2.595	6
7	17.086	.0585	.0311	.5311	32.172	1.883	1.565	2.947	7
8	25.629	.0390	.0203	.5203	49.258	1.922	1.675	3.220	8
9	38.443	.0260	.0134	.5134	74.887	1.948	1.760	3.428	9
10	57.665	.0173	.00882	.5088	113.330	1.965	1.824	3.584	10
11	86.498	.0116	.00585	.5058	170.995	1.977	1.871	3.699	11
12	129.746	.00771	.00388	.5039	257.493	1.985	1.907	3.784	12
13	194.620	.00514	.00258	.5026	387.239	1.990	1.933	3.846	13
14	291.929	.00343	.00172	.5017	581.859	1.993	1.952	3.890	14
15	437.894	.00228	.00114	.5011	873.788	1.995	1.966	3.922	15
16	656.841	.00152	.00076	.5008	1 311.7	1.997	1.976	3.945	16
17	985.261	.00101	.00051	.5005	1 968.5	1.998	1.983	3.961	17
18	1 477.9	.00068	.00034	.5003	2 953.8	1.999	1.988	3.973	18
19	2 216.8	.00045	.00023	.5002	4 431.7	1.999	1.991	3.981	19
20	3 325.3	.00030	.00015	.5002	6 648.5	1.999	1.994	3.987	20
21	4 987.9	.00020	.00010	.5001	9 973.8	2.000	1.996	3.991	21
22	7 481.8	.00013	.00007	.5001	14 961.7	2.000	1.997	3.994	22
23	11 222.7	.00009	.00004	.5000	22 443.5	2.000	1.998	3.996	23
24	16 834.1	.00006	.00003	.5000	33 666.2	2.000	1.999	3.997	24
25	25 251.2	.00004	.00002	.5000	50 500.3	2.000	1.999	3.998	25
26	37 876.8	.00003	.00001	.5000	75 751.5	2.000	1.999	3.999	26
27	56 815.1	.00002	.00001	.5000	113 628.3	2.000	2.000	3.999	27
28	85 222.7	.00001	.00001	.5000	170 443.4	2.000	2.000	3.999	28
29	127 834.0	.00001		.5000	255 666.1	2.000	2.000	4.000	29
30	191 751.1	.00001		.5000	383 500.1	2.000	2.000	4.000	30
31	287 626.6			.5000	575 251.2	2.000	2.000	4.000	31
32	431 439.9			.5000	862 877.8	2.000	2.000	4.000	32

Compound Interest Factors

	Single Payment		Uniform Payment Series				Arithmetic Gradient		
	Compound Amount Factor	Present Worth Factor	Sinking Fund Factor	Capital Recovery Factor	Compound Amount Factor	Present Worth Factor	Gradient Uniform Series	Gradient Present Worth	
n	Find *F* Given *P* *F/P*	Find *P* Given *F* *P/F*	Find *A* Given *F* *A/F*	Find *A* Given *P* *A/P*	Find *F* Given *A* *F/A*	Find *P* Given *A* *P/A*	Find *A* Given *G* *A/G*	Find *P* Given *G* *P/G*	**n**
1	1.600	.6250	1.0000	1.6000	1.000	0.625	0	0	**1**
2	2.560	.3906	.3846	.9846	2.600	1.016	0.385	0.391	**2**
3	4.096	.2441	.1938	.7938	5.160	1.260	0.698	0.879	**3**
4	6.554	.1526	.1080	.7080	9.256	1.412	0.946	1.337	**4**
5	10.486	.0954	.0633	.6633	15.810	1.508	1.140	1.718	**5**
6	16.777	.0596	.0380	.6380	26.295	1.567	1.286	2.016	**6**
7	26.844	.0373	.0232	.6232	43.073	1.605	1.396	2.240	**7**
8	42.950	.0233	.0143	.6143	69.916	1.628	1.476	2.403	**8**
9	68.719	.0146	.00886	.6089	112.866	1.642	1.534	2.519	**9**
10	109.951	.00909	.00551	.6055	181.585	1.652	1.575	2.601	**10**
11	175.922	.00568	.00343	.6034	291.536	1.657	1.604	2.658	**11**
12	281.475	.00355	.00214	.6021	467.458	1.661	1.624	2.697	**12**
13	450.360	.00222	.00134	.6013	748.933	1.663	1.638	2.724	**13**
14	720.576	.00139	.00083	.6008	1 199.3	1.664	1.647	2.742	**14**
15	1 152.9	.00087	.00052	.6005	1 919.9	1.665	1.654	2.754	**15**
16	1 844.7	.00054	.00033	.6003	3 072.8	1.666	1.658	2.762	**16**
17	2 951.5	.00034	.00020	.6002	4 917.5	1.666	1.661	2.767	**17**
18	4 722.4	.00021	.00013	.6001	7 868.9	1.666	1.663	2.771	**18**
19	7 555.8	.00013	.00008	.6011	12 591.3	1.666	1.664	2.773	**19**
20	12 089.3	.00008	.00005	.6000	20 147.1	1.667	1.665	2.775	**20**
21	19 342.8	.00005	.00003	.6000	32 236.3	1.667	1.666	2.776	**21**
22	30 948.5	.00003	.00002	.6000	51 579.2	1.667	1.666	2.777	**22**
23	49 517.6	.00002	.00001	.6000	82 527.6	1.667	1.666	2.777	**23**
24	79 228.1	.00001	.00001	.6000	132 045.2	1.667	1.666	2.777	**24**
25	126 765.0	.00001		.6000	211 273.4	1.667	1.666	2.777	**25**
26	202 824.0			.6000	338 038.4	1.667	1.667	2.778	**26**
27	324 518.4			.6000	540 862.4	1.667	1.667	2.778	**27**
28	519 229.5			.6000	865 380.9	1.667	1.667	2.778	**28**

Exam Files

Professors around the country have opened their exam files and revealed their examination problems and solutions. These are actual exam problems with the complete solutions prepared by the same professors who wrote the problems. **EXAM FILES** are currently available for these topics:

Calculus I
Calculus II
Calculus III
Chemistry
Circuit Analysis
College Algebra
Differential Equations
Dynamics
Fluid Mechanics
Linear Algebra
Materials Science
Mechanics of Materials
Organic Chemistry
Physics I—Mechanics
Physics II—Heat, Light, and Sound
Physics III—Electricity and Magnetism
Probability and Statistics
Statics
Thermodynamics

The **EXAM FILE** series also includes four engineering license review books:

Civil Engineering License Exam File
Electrical Engineering License Exam File
Engineer-In-Training Exam File
Mechanical Engineering License Exam File

The **EXAM FILES** for courses (about $14.45 to $21.45) and engineering licenses (about $22.45 to $36.45) may be ordered with a Visa or MasterCard by telephone.

Engineering Press/Maxway: (212) 947-6100

For a detailed description of the **EXAM FILES**, ask at your college or technical bookstore, or write to:

Engineering Press, Inc.
P.O. Box 1
San Jose, California 95103-0001